Rapid Assessment Program

19

RAP Bulletin of Biological Assessment

A Biological Assessment of the Aquatic Ecosystems of the Río Paraguay Basin, Alto Paraguay, Paraguay

Barry Chernoff, Philip W. Willink, and Jensen R. Montambault, Editors

Center for Applied Biodiversity Science (CABS)

Conservation International (CI)

The Field Museum

Museo Nacional de Historia Natural del Paraguay

Universidad Nacional de Asunción, Paraguay

RAP Bulletin of Biological Assessment is published by:
Conservation International
Center for Applied Biodiversity Science
Department of Conservation Biology
1919 M Street NW, Suite 600
Washington, DC 20036
USA
202-912-1000 tel
202-912-1030 fax
www.conservation.org
www.biodiversity science.org

Editors: Barry Chernoff, Philip W. Willink, and Jensen R. Montambault
Design/Production: Kim Meek
Maps: Mark Denil
Cover photographs: Kim Awbrey
Translations:
    Spanish: Gina Fognani and Jensen R. Montambault
    Portuguese: Yuri Leite
    Guaraní: Sara Musinsky

ISBN: 1-881173-40-2
Library of Congress Control Number: 2001087008

RAP Bulletin of Biological Assessment was formerly RAP Working Papers. Numbers 1–13 of this series were published under the previous series title.

Suggested citation: Chernoff, B., P.W. Willink, and J.R. Montambault (eds.). 2001. A biological assessment of the Río Paraguay basin, Alto Paraguay, Paraguay. RAP Bulletin of Biological Assessment 19. Conservation International, Washington, DC.

Funding for this AquaRAP study and publication was generously provided by The Rufford Foundation and The Giuliani Family Foundation.

Printed on recycled paper.

Preface

The Río Paraguay is 2,550 kilometers long and its basin covers 1,100,000 kilometers² —the second largest in South America. The Río Paraguay begins in the highlands of southwestern Brasil and subsequently flows south along the Pantanal, the world's largest wetland. The Río Paraguay is home to thousands of species of plants and animals, many of which are endemic to the Río Paraguay basin. At the same time, the Río Paraguay is the lifeblood of many communities of humans who depend upon the Río Paraguay for food, for transportation, for recreation, for their livelihoods, and for their cultures. Additionally, the Río Paraguay brings important nutrients for the soils with the incoming floods and removes salts that accumulate in flood-plain soils with the receding floodwaters. In short, the Río Paraguay and its river basin serve as critical habitat for the both humans and all other biological organisms upon which the human population often depends for existence.

Despite its size, the Río Paraguay is a relatively shallow river. This fact, coupled with its numerous meanders, means that the Río Paraguay is not easily or efficiently navigated by large ocean-going vessels. To remedy the navigational situation and to give land-locked Bolivia access to the sea via the Río Paraguay, the Hidrovia Project proposes to straighten and to deepen the Río Paraguay. In order to investigate the potential effects of Hidrovia upon the biodiversity and the aquatic habitats of the Río Paraguay, a team of 15 scientists, 2 logistical coordinators and one internet correspondent undertook an AquaRAP expedition to the Río Paraguay for three weeks in September 1997. The team was comprised of individuals from Paraguay, Brazil, Bolivia and the United States. The scientists were specialists in aquatic and terrestrial botany, decapod crustaceans, macro-invertebrates, ichthyology, and limnology—including water chemistry and plankton. The AquaRAP team contained both established scientists and students. The expedition focused primarily upon the biological and conservation value of the region and how

integrated solutions can work to preserve the maximum amount of biodiversity in the face of current and future threats. Additionally, the AquaRAP was featured on television in the United States in an episode of Bill Kurtis' *New Explorer's Series,* entitled "River in Peril."

The organization of this report begins with an executive summary, which includes a brief overview of the physical and terrestrial characteristics of the region, as well as summaries of the technical reports of the scientific disciplines, and concludes with recommendations for a conservation strategy. We then present the biological results from botany, limnology, decapod crustaceans and fishes. The last chapter highlights the commonalities among different floral or faunal elements in relation to geography and ecology. The disciplinary reports are written as scientific papers, each self-contained with its own literature citations for ease of use. After the scientific papers, we include a glossary of terms and then the appendices with much of the raw data.

We would like to explain our use of the terms *diversity* and *species richness*. In vernacular usage *diversity,* when used in the context of biological organisms, refers to the numbers of and variety of types of species, organisms, taxa, etc. In the ecological literature, *diversity* takes on a slightly different, more specific meaning, referring to the number of entities in combination with their relative abundances. In this volume *diversity* is used in two ways: (i) in the general, vernacular, sense we occasionally refer to the "*diversity* of organisms", meaning the number of and variety of organisms; but (ii) in the ecological sense, we refer to "low or high *diversity*", meaning "*diversity*" as calculated from a specific formula. Note that in the vernacular usage, *diversity* is never modified by high or low or used in a comparative way. From the ecological literature, the term *species richness* means the number of species. We use *species richness* in several chapters (e.g., Chapter 7) when referring to the number of species present in a habitat or river basin.

This report is intended for decision makers, environmental managers, governmental and non-governmental agencies, students and scientists. The novel information and analyses presented herein have two aims: (i) to present a compelling case and cogent strategy for conservation efforts within the region; and (ii) to provide scientific data and analyses that will stimulate future scientific research of this critical region. We have attempted in this volume not simply to present an inventory of the organisms that we encountered during our expedition but rather to use that information to evaluate conservation strategies under different scenarios of environmental threat (e.g., Chapter 7). We welcome comments and criticisms as we continue to evolve AquaRAP and the methods used for evaluating conservation strategies from biological data.

Barry Chernoff
Philip W. Willink
Jensen R. Montambault

Table of Contents

Participants and Contributors

Antonio Machado Allison
Instituto de Zoologia Tropical
Facultad de Ciéncias
Universidad Central de Venezuela
Caracas 1041-A, Venezuela
Email: amachado@strix.ciens.ucv.ve

Kim Awbrey (coordinator)
318 Yale Avenue
New Haven, CT 06515
Email: k.awbrey@att.net

Francisco Antonio R. Barbosa (limnology)
Universidade Federal de Minas Gerais, ICB
Departamento de Biologia Geral
Laboratório de Limnologia
Caixa Postal 486
CEP 30.161-970, Belo Horizonte, MG Brasil
Fax: 55-31-499-2567
Email: barbosa@mono.icb.ufmg.br

Hamilton Beltrán (terrestrial botany)
Museo de Historia Natural
Av. Arenales 1256
Apdo. 14-0434
Lima 14, Perú
Email: hamilton@musm.edu.pe

Barry Chernoff (ichthyology/editor)
Department of Zoology
Field Museum
1400 South Lake Shore Drive
Chicago, IL 60605 USA
Fax: 312-665-7932
Email: chernoff@fmnh.org

Karen Elizeche (botany)
Ciéncias Agrocias
Universidad Nacional de Asunción
Asunción, Paraguay

Glenda Fábregas (coordinator)
Center for Applied Biodiversity Science
Conservation International
1919 M Street, NW, Suite 600
Washington, DC 20036 USA
Fax: 202-912-0772
Email: g.fabregas@conservation.org

Bolívar Garcete (invertebrate zoology)
Facultad de Ciéncias Exactas y Naturales
Universidad Nacional de Asunción
Asunción, Paraguay

Arlo Hanlin Hemphill (web correspondent)
Fundación Jatun Sacha
Pasaje Eugenio Santillan N34-248 y Maurian
Urbanización Rumipamba
Casilla Postal #17-12-867
Quito, Ecuador
Fax: 593-2-432240
Email: arlo_hemphill@yahoo.com

Célio Magalhães (invertebrate zoology)
Instituto Nacional de Pesquisas da Amazonia
Coordenacao de Pesquisas em Biologia Aquatica
Caixa Postal 478
CEP: 69.011-970, Manaus, AM Brasil
Fax: 55-92-643-3146
Email: celiomag@inpa.gov.br

Darío Mandelburger (ichthyology)
Sección Ictiología
Museo Nacional de Historia Natural del Paraguay
Sucursal 1 Campus
Ciudad Universitaria, Central XI
San Lorenzo, Paraguay
Email: dmandelburger@sce.cnc.una.py

Mirta Medina (ichthyology)
Sección Ictiología
Museo Nacional de Historia Natural del Paraguay
Sucursal 1 Campus
Ciudad Universitaria, Central XI
San Lorenzo, Paraguay
Email: mnhnp@sce.cnc.una.py

María Fátima Mereles (aquatic botany)
Departmento de Botánica
Facultad de Ciencias Químicas
Universidad Nacional de Asunción
P.O. Box PY 11001-3291
Campus UNA—Paraguay
Fax: 595-21-58-5564
Email: Maríafmereles@yahoo.com

Jensen R. Montambault (editor)
Conservation International
1919 M Street, NW, Suite 600
Washington, DC 20036
Fax: 202-912-0772
Email: j.montambault@conservation.org

Gladys Arzamendia de Montiél (limnology)
Jubilada
Universidad Nacional de Asunción
Paraguay

Debra Moskovits (coordinator)
Environmental and Conservation Programs
Field Museum
1400 South Lake Shore Drive
Chicago, IL 60605-2496 USA
Fax: 312-665-7440
Email: dmoskovits@fmnh.org

Alexis Narvaez (ichthyology/genetics)
FACEN, Universidad Nacional de Asunción
P.O. Box PY 11001-3291
Campus UNA—Paraguay
Fax: 595-21-58-5564
Email: alexisnarvaez@hotmail.com

Marcos Callisto Faria Pereira (limnology)
Universidade Federal de Minas Gerais, ICB
Departmento de Biologia Geral
Laboratório de Limnologia
Caixa Postal 486
CEP 30.161-970 Belo Horizonte, MG Brasil

Jaime Sarmiento (ichthyology)
Museo Nacional de Historia Natural
Calle 26, Cota Cota
Casilla 8706
La Paz, Bolivia
Fax: 591-277-0876
Email: mnhn@mail.megalink.com

Mônica Solaga (aquatic botany)
Facultad de Ciéncias Exactas y Naturales
Universidad Nacional de Asunción
Asunción, Paraguay

Jørgen Thomsen (coordinator)
Conservation International
1919 M Street, NW, Suite 600
Washington, DC 20036
Email: j.thomsen@conservation.org

Mônica Toledo-Piza (ichthyology)
Departamento de Zoologia
Instituto de Biociências
Universidade de São Paulo
Caixa Postal 11461
CEP: 05422-970, São Paulo, SP Brasil
Email: mtpiza@usp.br

Juliana de Abreu Vianna (limnology)
Universidade Federal de Minas Gerais, ICB
Departmento de Biologia Geral
Laboratório de Limnologia
Caixa Postal 486
CEP 30.161-970 Belo Horizonte, MG Brasil

Sérgio Villanueva (ichthyology)
Facultad de Ciéncias Exactas y Naturales
Universidad Nacional de Asunción
Asunción, Paraguay

Philip W. Willink (editor)
Department of Zoology
Field Museum
1400 South Lake Shore Drive
Chicago, IL 60605 USA
Fax: 312-665-7932
Email: pwillink@fmnh.org

Pablo Moreno Souza Paula (limnology)
Universidade Federal de Minas Gerais, ICB
Departmento de Biologia Geral
Laboratório de Limnologia
Caixa Postal 486
CEP 30.161-970 Belo Horizonte, MG Brasil

Organizational Profiles

Conservation International

Conservation International (CI) is an international, nonprofit organization based in Washington, DC. CI believes that the Earth's natural heritage must be maintained if future generations are to thrive spiritually, culturally, and economically. Our mission is to conserve the Earth's living heritage, our global biodiversity, and to demonstrate that human societies are able to live harmoniously with nature.

Conservation International
1919 M Street, NW, Suite 600
Washington, DC 20036 USA
1-800-406-2306
202-912-0772 (fax)
http://www.conservation.org/
http://www.biodiversityscience.org/xp/CABS/research/rap
/aquarap/equaran.xml

Field Museum

The Field Museum (FMNH) is an educational institution concerned with the diversity and relationships in nature and among cultures. Combining the fields of Anthropology, Botany, Geology, Paleontology and Zoology, the Museum uses an interdisciplinary approach to increasing knowledge about the past, present, and future of the physical earth, its plants, animals, people, and their cultures. In doing so, it seeks to uncover the extent and character of biological and cultural diversity; similarities and interdependencies so that we may better understand, respect, and celebrate nature and other people. Its collections, public learning programs, and research are inseparably linked to serve a diverse public of varied ages, backgrounds, and knowledge.

Field Museum
1400 South Lake Shore Drive
Chicago, IL 60657 USA
312-922-9410
312-665-7932 (fax)
http://www.fieldmuseum.org/

Museo Nacional de Historia Natural del Paraguay

The Museo Nacional de Historia Natural del Paraguay is a young institution created in the 1980's, but has recorded a large part of the flora and fauna of Paraguay. It is a technical branch of the new Secretary of the Environment of Paraguay; therefore it oversees some of the activities corresponding to the investigation of the biodiversity of the country. With sections dedicated to ichthyology, ornithology, herpetology, mammalogy, botany, and environmental education, each area contains specialists that are trained for their tasks. The goal of the Museo Nacional de Historia Natural del Paraguay is to obtain the best information possible on the flora and fauna of the country in order to realize a sustainable use of the existing resources.

Museo Nacional de Historia Natural del Paraguay
Sucursal 1 Campus
Ciudad Universitaria, Central XI
San Lorenzo, Paraguay

Department of Botany, Faculty of Chemical Sciences, Universidad Nacional de Asunción, Asunción, Paraguay

The Department of Botany has a mission to produce studies in the areas of taxonomy, ecology, and ethnobotany, including the study of medicinal plants. The herbarium, which is the largest in Paraguay, contains 70,000 specimens. There is the Jardín de Aclimatación in which native forest species are cultivated. The department also publishes a botanical scientific journal, ROJASIANA.

The Faculty of Chemical Sciences carries out education and research at the highest level. On the Faculty, a number of investigators have been awarded the National Prize for Science (Paraguay) as well as honorable mentions.

The Universidad Nacional de Asunción was founded at the end of the 1900's and is the largest university within Paraguay. The University pursues education and scientific research in such diverse disciplines as the natural sciences, social sciences, and medical sciences. The main campus is located in the capitol, Asunción, while smaller affiliate campuses are located in other parts of the country.

Departmento de Botánica
Facultad de Ciencias Químicas
Universidad Nacional de Asunción
P.O. Box PY 11001-3291
Campus UNA—Paraguay
595-21-58-5564 (fax)

Acknowledgements

This AquaRAP expedition was a success due to the hard work of many individuals, and the first among them was Kim Awbrey. Kim, then AquaRAP Coordinator, used her vision and skills to overcome all hurdles. Darío and Roberto Tonañez expertly handled logistics during the expedition. We would like to thank Karen Elizeche (assistant terrestrial botanist, Ciéncias Agrocias-Universidad Nacional de Asunción), Bolívar Garcete (assistant invertebrate zoologist, Facultad de Ciencias Exactas y Naturales, Universidad Nacional de Asunción), Mônica Soloaga (assistant aquatic botanist, Facultad de Ciencias Exactas y Naturales, Universidad Nacional de Asunción), and Sérgio Villaneuva (assistant ichthyologist, Facultad de Ciencias Exactas y Naturales, Universidad Nacional de Asunción) for their help in conducting field work, as well as Coronel Carlos Giménez who was our pilot in Paraguay.

Mary Ann Rogers (Field Museum) and Kevin Swagel (Field Museum) provided invaluable assistance during the processing and cataloging of fishes in Chicago. Christina Canales (Field Museum) and Anna Lucia Benítez (Field Museum) entered data and conducted literature searches. Anna Chlumsky (Field Museum) formatted the botany tables in the appendices. The limnology team would like to thank Rosa María Menéndez (UNA) and Gladys Arza-mendia de Montiél (UNA) for information on zoo- and phytoplankton.

This report is published as part of Conservation International's series *RAP Bulletin of Biological Assessment,* and the publishing team would like to thank the following members for consistent high quality and hard work which makes these reports a reality: Leeanne E. Alonso and Anthony Rylands for excellent editing and suggestions; Gina Fognani, Adilson Salvador, and Fátima Mereles for assisting with the Spanish translation; Yuri Leite, Mônica Harris, and Erika Guimarães for assistance with the Portuguese translation; Sara Musinsky for coordinating translations in Guaraní; Mark Denil for creating the maps; John Musinsky for providing satellite imagery; and Debby Moskovitz for her assistance on photo identification.

The fieldwork and AquaRAP program was funded through the generosity of The Rufford Foundation. Equipment used in the field and laboratory to analyze data and produce the final report was provided by thoughtful gifts from the following: Comer Science and Education Foundation, Jay Fahn, Joan and Selma Goldstein, and David and Janet Shores. Innovative web broadcasts from the field were supported by assistance from Brevard Community College. The publication of this report was made possible by the generous support of The Giuliani Family Foundation. Lastly, we express our sincerest gratitude to John McCarter, Russ Mittermeier, and Peter Seligmann for their continuing support of the AquaRAP program.

Report at a Glance

A Biological Assessment of the Aquatic Ecosystems of the Río Paraguay Basin, Alto Paraguay, Paraguay

1) Dates of Study:
AquaRAP Expedition: September 4–18, 1997

2) Description of Location:
The AquaRAP expedition took place in the country of Paraguay and surveyed the section of the Río Paraguay between the Río Negro to the north, and the Río Aquidabán to the south. This area is located between two globally unique ecosystems, the Pantanal to the northeast, and the Chaco to the west. The Río Paraguay is a large river with low banks. The surrounding countryside is seasonally flooded, creating extensive wetlands and numerous small lakes that are often abandoned river channels. Portions of the Río Apa and Riacho La Paz tributary were also explored. The Río Apa is a medium-sized river with alternating rapids and pools. Its high banks limit the extent of flooding.

3) Reason for AquaRAP Study:
This region was the focus of an AquaRAP expedition because while the area is sparsely populated, poorly surveyed, and has experienced relatively little human disturbance, it is currently faced with a large development project: the Hidrovia Paraguay-Paraná. The Hidrovia Project proposes to dredge and channel extensive portions of the Río Paraguay watershed, altering water flow and the seasonal flooding cycles. These modifications could be disastrous to the aquatic ecosystems. Human population levels, cattle ranching, and agriculture in the region are also increasing. An assessment was necessary to document the aquatic diversity and the state of the aquatic ecosystems in order to determine sustainable conservation and development strategies for the region.

4) Major Results:
The northern-most section of the Río Paraguay, in the country of Paraguay, is currently in good health. Human population levels are low, as are the number of farms and cattle ranches. Fisheries are underutilized. The life-histories of aquatic organisms are highly dependent upon seasonal flooding of the Río Paraguay. The floods also create the high aquatic habitat heterogeneity observed during the expedition. The Río Apa is presently in good health. Water quality is high, human population levels are relatively low, and agriculture, cattle ranching, and logging are minimal. The lower Río Apa near the junction with the Río Paraguay is similar, but the upper Río Apa is ecologically distinct. Its steeper gradient and high sand banks place it apart. A distinct flora and fauna reflect this habitat separation.

5) Number of species:
Terrestrial plants:	198 species
Aquatic plants:	187 species
Shrimps and crabs:	13 species
Snails and mussels:	10 species
Chironomidae:	26 species
Aquatic insects (not including Chironomidae):	85 species
Other aquatic invertebrates:	10 orders
Fishes:	173 species

6) New records for Paraguay:
Aquatic plants:	*Phyllanthus fluitans* (Euphorbiaceae)
Shrimps and crabs:	*Pseudopalaemon* sp. (for Río Paraguay basin)
Fishes:	5 species

7) New species discovered:

Chironomidae:	Possibly many
Shrimps and Crabs:	Possibly 1 new shrimp, *Pseudopalaemon* sp.
Fishes:	Two species: *Triportheus* n. sp. A and *Heptapterus* n. sp.

8) Conservation Recommendations:

It is critical to prevent large- and small-scale disruptions of the seasonal flooding of the Río Paraguay, which would result in loss of wetlands and thereby devastate the aquatic biodiversity. Conservation of portions of both the Río Paraguay **and** the Río Apa is recommended. The area surveyed by the AquaRAP expedition is particularly attractive for the establishment of conservation areas because there are fewer people here than in the more heavily populated regions of the Río Paraguay to the south. Cooperation between Paraguay, Brasil, and Bolivia is encouraged since these drainages cross the country boundaries.

Informe a un vistazo (Español)

Una Evaluación Rápida de los Ecosistemas Acuáticos de la cuenca del Río Paraguay, Alto Paraguay, Paraguay.

1. Fecha de estudio

Expedición AquaRAP: 4–18 de septiembre, 1997.

2. Descripción del sitio:

La expedición AquaRAP se llevó a cabo en Paraguay; se inventarió la sección del Río Paraguay entre el Río Negro al norte y el Río Aquidabán más al sur. Esta área está ubicada entre dos ecosistemas que son únicos a nivel global: el Pantanal al noreste y el Chaco al oeste. El Río Paraguay es un río muy grande con barrancas relativamente bajas, dejando una importante área de inundación durante el periodo de aguas altas, en donde se forman humedales muy extensos y lagunas pequeñas pero numerosas, las que son originariamente cauces abandonados del río. Porciones del Río Apa y el tributario riacho La Paz también fueron investigados. El Río Apa es un río de tamaño medio que alterna entre rápidos y cursos de agua. Sus barrancas altas limitan la extensión de las inundaciones.

3. Razones para el estudio AquaRAP

La región fue el enfoque de una expedición RAP debido a que mientras la zona está muy poco habitada y ha tenido relativamente muy poca perturbación humana, está confrontada con un proyecto muy grande de desarrollo: la Hidrovía Paraguay-Paraná. El proyecto Hidrovía se propone dragar y canalizar porciones muy extensas de la cuenca del Río Paraguay, cambiando la corriente del agua y los ciclos temporales de inundación. Estas modificaciones pueden causar un desastre en el ecosistema acuático. Los niveles de población humana, la ganadería y la agricultura en la región también se encuentran en aumento. Una evaluación fue necesaria para documentar la diversidad acuática y así poder delinear una estrategia de conservación y desarrollo sostenible para la zona.

4. Resultados mayores

La sección más nórdica del Río Paraguay en Paraguay, todavía se encuentra en buen estado. Los niveles de población humana son aún bajos, al igual que el número de haciendas para la ganadería. Los recursos de pesqueros son muy poco utilizados por la población. La vida natural de los organismos acuáticos depende de las inundaciones del Río Paraguay. Las inundaciones producen también la alta heterogeneidad de hábitats acuáticas que se observó durante la expedición.

El Río Apa se encuentra en buen estado. La calidad del agua está buena y el nivel de la población humana está relativamente bajo; la agricultura, la ganadería y la deforestación son mínimas. El bajo Río Apa, en las cercanías de su desembocadura con el Río Paraguay es bastante parecido a éste; sin embargo, el Río Apa más arriba tiene una ecología distinta. Su grado de elevación más marcado y sus altos barrancos de arena lo distinguen. Una flora y fauna distinta a la del Río Paraguay refleja esta separación de hábitat.

5. Número de especies

Plantas terrestres:	198 especies
Plantas acuáticas:	187 especies
Camarones y cangrejos:	13 especies
Caracoles y mejillones:	10 especies
Chironomidae:	26 especies
Insectos acuáticos: (sin contar con Chironomidae)	85 especies
Otros invertebrados acuáticos:	10 ordenes
Peces:	173 especies

6. Nuevos registros para el Paraguay

Plantas acuáticas:	*Phyllanthus fluitans* (Euphorbiaceae)
Camarones y cangrejos:	*Pseudopalaemon* sp. (para la cuenca del Río Paraguay)
Peces:	5 especies

7. Nuevas especies descubiertas

Chironomidae:	Posiblemente muchos
Camarones y cangrejos:	Posiblemente 1 camarón, *Pseudopalaemon* sp.
Peces:	2 especies: *Triportheus* n. sp. A y *Heptapterus* n. sp.

8. Recomendaciones para la Conservación

Es importante prevenir las perturbaciones causadas por las inundaciones temporarias del Río Paraguay a escalas más amplias y restringidas, porque la pérdida de los humedales devastaría la biodiversidad acuática. La conservación de porciones en ambos Ríos (Paraguay y Apa) es una resultante de la expedición. Las áreas estudiadas por la expedición AquaRAP fueron muy apropiadas para establecer áreas conservadas, debido a que hay menos población en la porción del Río Alto Paraguay que en las áreas más al sur del mismo, las que se encuentran ya más pobladas. La cooperación entre el Paraguay, Brasil, y Bolivia debe ser alentada porque esta cuenca cruza la frontera de los tres países.

Resumo do Relatório (Português)

Levantamento Biológico dos Ecossistemas Aquáticos da Bacia do Rio Paraguai, Paraguai

1) Período do Estudo:
Expedição AquaRAP: 4 a 18 de setembro de 1997.

2) Descrição da Área:
A expedição AquaRAP foi realizada no Paraguai e explorou o trecho do Rio Paraguai entre os rios Negro ao norte e Aquidabán ao sul. Essa área se localiza entre dois ecossistemas únicos, o Pantanal ao nordeste e o Chaco ao oeste. O Rio Paraguai é um rio de grande porte e com margens baixas. A região é sazonalmente inundada, criando extensas áreas alagadas e um grande número de pequenos lagos que muitas vezes são braços mortos de rio. Partes do Rio Apa e do tributário Riacho de La Paz também foram exploradas. O Rio Apa é um rio de médio porte que alterna corredeiras e remansos com poços mais profundos. Suas margens altas limitam a extensão das cheias e alagados marginais.

3) Justificativa para o AquaRAP:
Essa região foi enfocada pela expedição AquaRAP por ser esparsamente povoada e pouco pesquisada, ter sofrido distúrbios antrópicos relativamente pequenos e estar sujeita a um projeto de desenvolvimento: a Hidrovia Paraguai-Paraná. O Projeto Hidrovia pretende dragar e canalizar grandes porções da bacia do Rio Paraguai, alterando o fluxo de água e os ciclos sazonais de cheias. Essas modificações poderão ser desastrosas para os ecossistemas aquáticos. A população humana, assim como a pecuária e a agricultura têm aumentado na região. Um estudo documentando a diversidade aquática e as condições do ecossistema aquático tendo como objetivo determinar estratégias de conservação e desenvolvimento sustentáveis para a região torna-se, portanto, fundamentalmente necessário.

4) Principais Resultados:
O extremo nordeste do Rio Paraguai, em território Paraguaio, encontra-se atualmente em boas condições ambientais. Os níveis populacionais humanos são baixos, assim como o número de fazendas agrícolas e pecuárias. A pesca é sub-utilizada. A biologia dos organismos aquáticos depende muito das inundações sazonais do Rio Paraguai. Essas cheias também criam grande heterogeneidade de habitats aquáticos observada durante a expedição.

O Rio Apa também se encontra em boas condições ambientais. A qualidade da água é alta, os níveis populacionais humanos são relativamente baixos e a agricultura, pecuária e retirada de madeira são mínimas. O baixo Rio Apa é similar ao Rio Paraguai, mas a parte alta do Rio Apa é ecologicamente distinta. Seu relevo íngreme e bancos de areia altos são característicos, fazendo com que sua fauna e flora sejam distintas.

5) Número de Espécies:

Plantas terrestres:	198 espécies
Plantas aquáticas:	187 espécies
Camarões e caranguejos:	13 espécies
Caramujos e mexilhões:	10 espécies
Chironomidae:	26 espécies
Insetos aquáticos (excluindo Chironomidae):	85 espécies
Outros invertebrados aquáticos:	10 ordens
Peixes:	173 espécies

6) Novos registros para o Paraguai:

Plantas aquáticas:	*Phyllanthus fluitans* (Euphorbiaceae)
Camarões e caranguejos:	*Pseudopalaemon* sp. (para a bacia do Rio Paraguai)
Peixes:	5 espécies

7) Espécies novas descobertas:

Chironomidae:	Provavelmente muitas
Camarões e caranguejos:	Provavelmente 1 novo camarão, *Pseudopalaemon* sp.
Peixes:	2 espécies: *Triportheus* n. sp. A e *Heptapterus n.* sp.

8) Recomendações para conservação:

É fundamental prevenir distúrbios em pequena e grande escala das inundações sazonais do Rio Paraguai que podem resultar na perda de áreas inundáveis e, portanto, devastação da biodiversidade aquática. Recomenda-se a conservação de partes do Rio Paraguai e do Rio Apa. A área pesquisada pela expedição AquaRAP é particularmente atraente para a implantação de unidades de conservação, pois a população humana nesse local é menor do que aquela nas regiões do Rio Paraguai situadas mais ao sul, que são densamente povoadas. Deve ser estabelecida uma cooperação entre o Paraguai, Brasil e Bolívia, pois esses cursos d'água cruzam as fronteiras desses países.

Mombe'u (Guaraní)

Petei Techauka Pya'e Ysyry Paraguai Pe Gua

AquaRap guata: Septiembre 4–18, 1997

Tenda:

AquaRap guata o che chapo Paraguai pe; o je hecha ha o jei papa oiva Ysyry Paraguai pe, tenda o pytaba mokôi veva mbytépe, Ysyry Negro yvate vo ha Ysyry Aquidabán yvygotyo. Ko tenda opyta mokôi tekoha apytépe ndaipóri va haecueraicha gua mamové ko yvy ári: Pantanal peteî yke pe yvate-kuarahyreike pe (al noreste) ha Chaco kuarahyresê (al oeste).

Ysyry Paraguai tucha ha yby tenda ijerere ikarape, upeare petei tenda guasu o jagarrapa ysyry oky iterei vove, avei tuchukua ha heta yno'ô asâi rupi ysyry taperepe. Avei Ysyry Apa pehengue ha Ysyry La Paz oñe kundaha. Ysyry Apa tuichakue vy oñe moambue va pya'e ha mbegue syry pe. Yby tenda ijerere i yvate gui o choco ysyry pono isarambi ha ituchu kua.

Maerâpa oñembo'euka AquaRap

Oje japo RAP guata Ko tenda pe oguerekojaguere poka oikova pype ha no ñembyai byi guiteriei ko'a§apeve, ko'á§a o ñembo rovái peteî jejaporâ tuicha iterei va: Tape ysyry Paraguai-Parana (Hidrovía Paraguay-Paraná). Jejaporâ Tape Ysyry jo'o se ha oi pe'a se heta pehengue oî va Ysyry Paragui gu″pe, moambue ysyry syry ha y guasu ára, ko moambue ikatu o gueru geta mbae bai tekoha ysyry gu″pe guape. Yvypóra, ñemitÿ kuaa, (ganadería) avei o kakuaa. Peteî je hecha ha je papa te kotee vacue o je kuaa ha§ua mbae pa oî y gu″pe ha upea rupi cha juhu ha§ua peteî moatâ tekoha ñangareko ha táva kakuaa.

Guata rire o je kuaa

Ysyry Paragui yvate yto, Paragui pe, i poraiterei gueteri. Tenda tekove gua ipoka, avei nda hetai gueteri yby vaka tenda. Tetâ gua ndai porui guasu pira. Tekove y gu″pe gua oikoteve y muasâi yv″ari ysyry pukukuévo. Y muasâi rupi oiko ve heta y tekove renda o je hecha vakue guata hape.

Ysyry Apa iporâ gueteri. Y ipoti ha tenda gua ipoka gueteri; ñemitÿ kuaa, tenda vaka ioko ha, ha yvyra takâ ñe kytî ipoka. Ysyry Apa ha Ysyry Paraguai o che choapy hape ypy rupi upeicha avei; ko'"te i mboyp"iri Ysyry Apa oguereko tekoha ndo choguaiva maavea pe. Tenda Ysyry pukukuévo yvate ha i yvyku'i. Peteî ka'aguy ha ka'aguy mymba ndo choguai Ysyry Paraguai gua pe, u peare jai kuaa oî ha mokôi tekoha.

Papaha:

Plantas terrestres:	198 especies
Plantas acuáticas:	187 especies
Camarones y cangrejos:	13 especies
Caracoles y mejillones:	10 especies
Chironomidae:	26 especies
Insectos acuáticos:	85 especies
(sin contar con Chironomidae)	
Otros invertebrados acuáticos:	10 ordenes
Peces:	173 especies

6. Nuevos registros para el Paraguay

Plantas acuáticas:	*Phyllanthus fluitans* (Euphorbiaceae)
Camarones y cangrejos:	*Pseudopalaemon* sp. (para la cuenca del Río Paraguay)
Peces:	5 especies

7. Nuevas especies discubiertas

Chironomidae:	Posiblemente muchos
Camarones y cangrejos:	Posiblemente 1 camarón, *Pseudopalaemon* sp.

Peces: 2 especies: *Triportheus* n. sp. A
                          y *Heptapterus* n. sp.

Porânguereko tekoha ñangareko ra

Tekotevê jai kuaa ha ja choko opaite mbae o mbo marâ va
(guaiva) Ysyry Paraguai muasâi yby ari, o ñehundi ro yby
ykue, tuyu kua, oñehundita avei heta y mymba. Tekoha ñangareko mokôi veva Ysyry pe Paraguai ha Apa ñande mboe
AquaRap jeguata. Tenda AquaRap o kundaha haguepe i porâ
o je ñangareko agua, i nandi ve gui yvypóra gui Alto Paragui
ngotyo, tenda opytava yvygotyo ve o guereko hetave tekove
kuera. Tekotevê Paraguai, Brasil ha Bolivia o mba'apo vara
oñondivepa o ñangareko ha gua ha o ñe mo porâve hague ko
Ysyry ohasava mbohapy veva ypyrupi.

Executive Summary

Introduction

The Río Paraguay is a large, meandering river that, when merged with the Río Paraná, forms the second largest river basin in South America (Map 1). The size of the watershed belies the fact that the Río Paraguay is a relatively shallow river, discharging its share of only 314 kilometers^3 of water per year for the entire Paraguay-Paraná basin. This is low compared to the Río Orinoco basin which discharges more than four times that value (Gleick, 1998). The Río Paraguay, which has its headwaters in the Planalto do Mato Grosso in Brasil, flows through the Pantanal wetland and along the eastern border of the Chaco, dividing the country of Paraguay (Map 1). The Río Paraguay is a critical aquatic resource for the people of Paraguay and Brasil, not only because of its rich biodiversity, but also because of human dependence upon the fisheries resources and freshwater for consumption and agriculture.

The biodiversity of the Río Paraguay represents an interesting transition between tropical Amazonian rivers to the north and temperate habitats to the south. Geological disturbances over the past 90 million years have provided opportunities for rivers and their inhabitants to switch back and forth between the Amazon and Paraguay basins (Heckman, 1998:31–32; Lundberg et al., 1998). The similarity of the aquatic flora and fauna to that of the Amazon is believed to be due to headwater exchanges between the Ríos Guaporé and Arinos (both of the Amazon) and the headwaters of the Río Paraguay during flooding (Pearson, 1937; Roberts, 1972; Lowe-McConnell, 1975:47,85; Bonetto, 1986a,b). Past and present aquatic connections help explain the presence of species in Paraguay which are commonly found along the Atlantic coast or to the south in Argentina and Uruguay. With the addition of endemism, the assemblages of organisms in the Río Paraguay are unique (Heckman, 1998:31–33).

Characterization of the flora and fauna of the Río Paraguay is critical to document the aquatic diversity and the state of the aquatic ecosystems in order to determine sustainable conservation and development strategies for the region. Threats can take many shapes, but the principal cause for concern is the planned Hidrovia (Paraguay-Paraná) Project. The Hidrovia ("hidrovia" means waterway) is a project designed to make 3,400 kilometers (2,100 miles) of the Río Paraguay and Río Paraná navigable to barges and large ships by dredging, straightening bends, digging new channels, and destroying rock outcroppings. The area affected will extend from Cáceres, Brasil on the Río Paraguay, downstream to the mouth of the Río Paraná at the Atlantic Ocean (Map 1). Many of the river modifications would be in the Pantanal (Lourival et al., 1999; Switkes, 1996a,b,c; WWF, 2001) with the intention of facilitating the year-round transportation of principally soybeans and iron ore. Argentina, Uruguay, Paraguay, Bolivia, and Brasil are involved in the project (Moore, 1997; Heckman, 1998:497).

Unintended environmental consequences of the project could include draining and drying of large sections of the Pantanal wetlands, decreased flooding in the upper parts of the Río Paraguay watershed, and increased flooding downstream (Switkes, 1996b,c, 1997). Salinization of certain tributaries, contamination of drinking waters, and increased risk of oil spills due to increased river traffic are probable consequences to this project. The economic benefits are believed to have been overestimated because existing road traffic and potential competition with railroads were not taken into account (Switkes, 1996c, 1997; Moore, 1997; Heckman, 1998:497–498).

Habitat destruction and conversion is not restricted to the Hidrovia Project. Population levels in Paraguay are increasing, along with agriculture, forestry, and cattle ranching. These activities are usually most intense near settlements and transportation routes, such as roads and waterways. Agriculture and cattle ranching involve burning and clearing of large tracts of land. When unrestrained and conducted over long periods of time, this modification of the habitat will jeopardize the biodiversity in a given area and increase erosion.

The AquaRAP Expedition

In response to the Hidrovia Project and other threats, a multidisciplinary, multinational team of scientists undertook an Aquatic Rapid Assessment Program (AquaRAP) expedition along the Río Paraguay with the goals of: (i) describing the biological and environmental aspects of the aquatic ecosystems before the implementation of any large scale modifications, and (ii) determining, if possible, the potential changes to the aquatic resources in response to perceived threats. The expedition took place between September 4 and 18, 1997, and surveyed the section of the Río Paraguay between the Río Negro, to the north, and the Río Aquidabán, to the south (Map 1). Portions of the Río Apa and Riacho La Paz tributary were also investigated. To allow for comparisons, the area surveyed was divided into five regions: Río Negro, Upper Río Paraguay (upstream from Cerritos Pão de Açúcar, approximately 21° 26'S, 57° 55' W), Lower Río Paraguay (downstream from Cerritos Pão de Açúcar), Río Apa, and Riacho La Paz (Map 1). Terrestrial vegetation, aquatic vegetation, physical and chemical water characteristics, plankton, aquatic macro-invertebrates, and fishes were assessed. These inventories, in light of the proposed Hidrovia Project and other factors, will be used as a basis for conservation and future research recommendations in the area.

The South American Aquatic Rapid Assessment Program (AquaRAP) is a multinational, multidisciplinary program devoted to identifying conservation priorities and sustainable management opportunities for freshwater ecosystems in Latin America. AquaRAP's mission is to assess the biological and conservation value of tropical freshwater ecosystems through rapid inventories, and to report the information quickly to local policy makers, conservationists, scientists, and international funding agencies. AquaRAP is a collaborative program managed by Conservation International and the Field Museum.

At the core of AquaRAP is an international steering committee composed of scientists and conservationists from seven countries (Bolivia, Brasil, Ecuador, Paraguay, Perú, Venezuela, and the United States). The steering committee oversees the protocols for rapid assessment and the assignment of priority sites for rapid surveys. AquaRAP expeditions, which involve major collaboration with host country scientists, also promote international exchange of information and training opportunities. Information gathered in AquaRAP expeditions is released through Conservation International's *RAP Bulletin of Biological Assessment* designed for local decision-makers, politicians, leaders, and conserva-

tionists, who can set conservation priorities and guide action through funding in the region.

An assessment of the current biodiversity in the Río Paraguay and Río Apa was critical before large scale modifications were made to the region. This part of Paraguay has been woefully under-sampled. Most of the collections made in the past were done in the vicinity of large population centers. Additional information was necessary to allow for the recognition of management opportunities. The following section summarizes the observations made during the AquaRAP expedition. Please see the chapters themselves for more detailed information. Chapter summaries appear after the abovementioned synthesis, which is then followed by recommended conservation and research initiatives.

Overall Summary

The AquaRAP expedition increased our knowledge of the region, which, prior to the expedition, was limited to forestry inventories, hydrologic and salinization studies (Río Verde), some littoral vegetation studies, and sporadic fish collections. This expedition compiled concurrent botanical, limnological, invertebrate, and ichthyological data. Integration of this information allowed us to assemble a comprehensive view of the entire aquatic ecosystem. Based on this comprehensive view, we make the observations and recommendations below.

Río Paraguay

The northern-most section of the Río Paraguay, in the country of Paraguay, is located between two major habitat types found nowhere else in the world, the Pantanal and the Chaco. The Pantanal is the world's largest wetland, renowned for its tremendous flocks of birds and large numbers of caiman, capybara, deer, anteaters, and other large fauna. The open terrain makes it an animal-watcher's paradise. The Chaco is a vast plain of rugged beauty dominated by grasses and scrub-brush. Water is scarce, being replenished only by the seasonal rains and floods. The Chaco flora and fauna, which includes jaguars, rheas, deer, maned wolves, anacondas and other snakes such as boas and rattlesnakes, anteaters, tapirs, peccaries including the taguá (*Catagonus wagneri*— previously thought to have been extinct) (Sowls, 1997), and groups of birds, many migratory, is similar to those of the Pantanal. These creatures have become adapted to the harsh conditions and are surprisingly numerous. Numbers and diversity of organisms are higher along rivers and marshes (Weil et al., 1984:17–19). The global uniqueness of the Pantanal and Chaco and their dependence on specific hydrological phenomena make them very amenable to aquatic conservation initiatives, provided that these regions have not been damaged severely by human activities.

The area studied by the AquaRAP expedition is currently in good health. The local people are heavily dependent upon the water for survival, but human population levels and the number of farms and cattle ranches are low. Productivity is high, with aquatic macrophytes responsible for the majority of primary production and decomposition which bring nutrients to the river. Fisheries are underutilized. No direct contamination of the water was noted. Unfortunately, this situation is likely to change.

Population levels are expected to rise. There is currently room for development in the region, but unregulated growth will cause unnecessary long-term problems. Impact studies are critical to determine how to maintain farms and fisheries at sustainable levels. Additional agricultural farms and cattle ranches will increase erosion and result in the loss of valuable species, as is already occurring near Bahía Negra. Commercial and residential water pollution, with subsequent eutrophication and health problems, is likely. Precautions, such as maintaining riparian vegetation to limit erosion, and waste treatment of used water, must be taken to avoid these negative outcomes. Commercial and tourist fishing pressure may increase, and the reproductive and growth rates of the fishes involved should be considered when drafting legislation. If population dynamics are ignored, then it is only a matter of time before fisheries collapse. In general, more work needs to be done on the ecology of the region to determine sustainable levels of agriculture and fisheries.

Perhaps the greatest threat comes from a proposed project, the Hidrovia Paraguay-Paraná, designed to make the Río Paraguay navigable to large, ocean-going ships. The exact impact of the Hidrovia Project is unknown (see pp. 50). Portions of the Pantanal wetlands near the headwaters of the Río Paraguay are anticipated to dry out, and flooding is expected to increase downstream (Lourival et al., 1999; Switkes, 1996b,c, 1997; WWF, 2001). The extent of the hydrological impact on the study area is difficult to predict, but either consequence will alter the current distribution and composition of aquatic communities. The fact that the full extent of the damage from dredging and channeling cannot, or is not being predicted, is reason enough for concern.

Decreased flooding and the resulting loss of wetlands will be devastating to aquatic communities. The dominant ecological phenomenon of the region is the seasonal flooding, with local aquatic and semi-aquatic organisms adapted to the periodic inundations. These inundations redistribute nutrients, facilitating the growth of vegetation in wetlands. Many plants also depend on floods to disperse their seeds (Goulding, 1980). Rising water levels are known to serve as cues for the reproductive migrations of several commercially-valuable species of fish and, for many organisms, the majority of their annual growth occurs during periods of flooding (Lowe-McConnell, 1975; Goulding, 1980; Bonetto, 1986b). Inundated areas act as nurseries and shelters for many species of fishes and invertebrates (Bonetto, 1986b; Nakatani et al.,

1997). Disruption of flooding will dramatically decrease habitat heterogeneity, biodiversity, and biomass.

Effects of the Hidrovia Paraguay-Paraná will not be limited to the aquatic ecosystem. The number of individual fishes will decrease, spelling disaster for fisheries, as well as birds, mammals, and reptiles which rely on aquatic macro-invertebrates and fishes for food. Downstream, increased flooding will destroy existing cropland and force the resettlement of people living too close to the river.

Río Apa

The Río Apa is presently in good health. Water quality is high and human population levels are relatively low. Agriculture, cattle ranching, and logging are present in the area. The river provides clean water and food for local communities and for various birds, reptiles, and mammals.

Although the lower Río Apa is similar to the Río Paraguay, the upper part of the river is ecologically distinct, with a steeper gradient and sand banks. A different flora and fauna reflect this habitat distinctness. Efforts should be made to set aside areas in the upper and lower reaches of the river. By saving representative samples of the Río Apa and the Río Paraguay, the habitat heterogeneity and biodiversity of the entire region can be maintained.

With its steep banks and unconsolidated soil, the Río Apa is particularly susceptible to erosion. Riparian vegetation is currently holding much of the soil in place. This natural and effective anti-erosion measure is threatened by logging and slash-and-burn agriculture, both of which remove the much needed vegetation.

Erosion can cause multiple problems. One is the loss of valuable topsoil, eventually resulting in decreased crop yields. Agriculture will then be sustained only with the use of fertilizers. Fertilizers will be washed into the river and eutrophication will occur. Another problem is the silting of the river. Too much erosion will cause the Río Apa to become turbid, no longer possessing unique habitats, and subsequently reducing the regional habitat heterogeneity and biodiversity.

Lack of regulation is the primary reason that slash-and-burn agriculture and logging are threats. If steps are not taken to limit habitat destruction, particularly along the river, long term siltation and erosion problems will occur. Regulations concerning the types of trees that can be harvested will also ensure the long-term economic success of the region and prevent the extinction of valuable timber species such as *Amburana caerensis*. With an appropriate management plan, it may be possible to sustain productivity while still maintaining the biological integrity of the aquatic ecosystem.

The Hidrovia Project will impact the lower Río Apa in a manner similar to that discussed above for the Río Paraguay. The hydrological consequences on the upper Río Apa are unclear because it is elevated above the Río Paraguay. There is a possibility that groundwater will drain more quickly, reducing the river size without groundwater input. Under this scenario, populations of aquatic organisms will decrease, but habitat types and diversity will not be expected to change unless the volume of flow drops substantially. Increased flooding is not anticipated in the upper Río Apa.

Chapter Summaries

Geographic Overview

The Río Paraguay is 2,550 kilometers (1,580 miles) long and its basin covers 1,100,000 kilometers² (420,000 miles²) (Bonetto, 1986a). It begins in the highlands of Planalto do Mato Grosso, referred to as Chapada dos Parecis, in southwestern Brasil, and subsequently flows south along the Pantanal, the world's largest wetland (Map 1). After exiting the Pantanal, the Río Paraguay splits the country of Paraguay into western and eastern halves before merging with the Río Paraná. Together, the Ríos Paraguay and Paraná form the second largest watershed in South America.

Though water levels in the rivers depend on the amount of rainfall, the relationship is not quite so simple. In the Pantanal, much of the rain falls during the summer months, but the low gradient and flooding of the wetlands maintains the water in Brasil. This water is slowly released, pulsing southward. According to Ferreira et al. (1994), peak water levels occur, on average, in the Río Paraguay basin as follows: in the northern Pantanal—February to March; at Corumbá—April to May; and at the mouth of the Río Apa—July to August. This hydrological phenomenon tempers the magnitude of downstream flooding. Even though water levels can change up to 4 meters (13 feet) in the Brasilian Pantanal, the Río Paraguay rises only 1.5 meters (5 feet) at Asunción (Weil et al., 1984:15).

The Río Paraguay, in northern Paraguay, is a large, meandering river flowing through relatively flat terrain. Water level fluctuations and the flat terrain result in widespread flooding and frequent changes in the course of the river. Oxbow lakes and abandoned channels are common, and the river's banks are lined with forests and palm stands. These are routinely flooded for several months each year. Large masses of floating vegetation (e.g., *Eichhornia* sp., *camalotes*), known as floating meadows, are common. Rock outcroppings near Cerritos Pão de Açúcar form a natural bottleneck, slowing the free flow of water downstream.

The Río Apa differs from the Río Paraguay in that it is a smaller river with cascades and waterfalls. The bottom is predominantly sand and the banks are high and well defined. Accumulations of tree trunks and branches are common. Habitat diversity in the Río Apa is lower than in the Río Paraguay, and water levels can change rapidly in response to rains.

The Riacho La Paz can be described as a headwater stream with a course that alternates between riffles and pools.

The river bottom is sand and rock, and water levels here, like in the Río Apa, can change rapidly in response to rains. Overall habitat diversity is low.

Terrestrial Vegetation

Twelve variable transects were established to assess the terrestrial vegetation along the Río Paraguay and the Río Apa (Map 2). General collections were also made of plants not encountered on the transects. A total of 198 plant species were identified. There were five basic types of habitats along the Río Paraguay: seasonally inundated palm stands of *Copernicia alba*, non-flooded palm stands of *C. alba*, modified forests of *Schinopsis balansae*, forested hills, and Chaco forests.

Seasonally inundated palm stands are very common along the Upper Río Paraguay. They are dominated by *C. alba*, but shrubs, lianas, and grasses are also abundant. The non-flooded palm stands of *C. alba* are often burned and converted into pasture. The result is a modified palm stand with abundant pioneer shrubs which modify this type of formation, such as: *Acacia caven*, *Prosopis ruscifolia*, and *Mimosa pellita*, among others. The modified forests of *S. balansae* were composed of a shrub-like layer of *Coccoloba* aff. *guaranitica*, *Tabebuia nodosa*, *Ziziphus mistol*, *Pseudobombax* sp., *Copernicia alba*, *Caesalpinia paraguariensis*, and *Prosopis ruscifolia*.

The forested hills are out of reach of the floods and the soil drains rapidly. The flora is therefore similar to that found in areas with little water. Trees tend to be small and grasses ephemeral. Chaco forests are similar, with trees reaching 9 to 12 meters in height. The undergrowth is dominated by spiny shrubs and grasses.

There are three basic types of habitats along the Río Apa: riparian vegetation, semi-deciduous forests, and cerradão. The riparian vegetation borders the Río Apa and is flooded during high waters. The semi-deciduous forests can be divided into three stories: the highest level is composed of trees 25 to 30 meters in height; the intermediate level is composed of trees between 15–20 meters in height and made up of *Croton urucurana*, *Trichilia* sp., and *Astronium fraxinifolium;* and the understory is made of a few bushes up to 1 meter in height forming thickets. Many of the species are deciduous, but during the rain events, this habitat would resemble a rain forest. Cerradão formations are furthest from the river, where larger trees are uncommon.

Overall, the terrestrial vegetation adjacent to the rivers is largely influenced by the fluctuations in water levels and the seasonality of the rains. Burning and habitat conversion, principally for cattle ranching, have had a significant impact on the region. Evidence for this is that pristine vegetation was not encountered during the expedition. Excess deforestation is accelerating erosion, resulting in major sedimentation of the Río Paraguay.

Aquatic Vegetation

Aquatic vegetation was sampled at 31 localities along the Río Paraguay and seven localities along the Río Apa (Map 2). A total of 147 species were identified along the Río Paraguay and 50 species along the Río Apa, for a grand total of 187 aquatic plant species.

The flooded banks of the Río Paraguay were dominated by a "mosaic of vegetation." The components of this "mosaic" were forests of *Schinopsis balansae,* savanna-palm stands of *Copernicia alba,* flooded lowlands, and shoreline vegetation. In the Río Paraguay, we focused on the shoreline vegetation. This vegetation was subdivided into semi-lotic, shoreline, swamp, and sand bank habitats. Species were categorized according to their life style: epiphytic, free-floating, liana, marsh, rooted semi-submergent, or submergent.

Habitats with slow-flowing water are found where the river overflowed its banks. These habitats are much more diverse than one would expect by chance, and probably account for the majority of aquatic floral diversity within the study area. This high diversity may be related to the dynamics of the water levels and nutrient cycles. Shoreline habitats are relatively poor in aquatic vegetation, possibly because the constant movement of water hinders the growth of many species.

Swamps are highly variable. Some, with little sediment, could better be described as "floating meadows". Others, usually in areas sheltered from the wind, currents and in more advanced stages of development, possess an appreciable amount of substrate. The number of species in a given area is roughly proportional to the amount of substrate.

Sand bank habitats are generally less rich in species, probably due to the poor quality of nutrients in their soils. However, the plant species encountered are very distinctive and characteristic of these sites.

The aquatic floral diversity of the Río Apa is much lower than that of the Río Paraguay. This is probably due to the high current velocity, lack of habitat diversity, preponderance of sand, and large numbers of trunks and branches throughout the channel. The more diverse environments are found in some of the tributaries.

Water Quality and Benthic Macro-invertebrates

Water quality was assessed at 35 localities (Fig. 4.1). The waters of the Río Paraguay were generally slightly acidic (pH 6.0–6.5), with low oxygen levels (<6.0 mg/L), low electrical conductivity (60–100 µS/cm), and temperatures ranging between 24–27° C. The waters of the Río Apa were neutral to slightly alkaline (pH 7.27–7.96), with variable levels of dissolved oxygen (4.2–7.13 mg/L), high electrical conductivity (163–376 µS/cm), and temperatures ranging between 22–29° C.

The phytoplankton communities were relatively diverse, and included representatives several algae groups, including Chlorophyta, Euglenophyta, Chrysophyta, Bacillariophyta, and Cyanophyta. Benthic macro-invertebrates were collected at 33 sites. Chironomidae (Diptera) larvae and Oligochaeta were the dominant groups at 27 of the sites, representing 52% and 35% of the recorded organisms. Odonata, Trichoptera, Ephemeroptera, Chaoboridae, Ceratopogonidae, Corixidae, Conchostraca, Planaria, Nematoda, Hirudinea, Copepoda, Ostracoda, Bivalvia, and Gastropoda were collected less frequently.

Among the Chironomidae larvae, 26 taxa were identified. Some may be undescribed species. The diversity of benthic macro-invertebrates is high compared to other watersheds in South America. The majority of the genera are typical for herbaceous marshes, ponds, lakes, and slow moving portions of streams and rivers. Areas rich in decomposing vegetation exhibited a low diversity.

Overall, there was little evidence for direct contamination of the waters. The only exceptions were near Bahía Negra, where the burning and clearing of land could result in erosion and eutrophication.

Macro-invertebrates

The aquatic macro-invertebrate fauna was surveyed at 54 collecting stations along the upper course of the Río Paraguay, between the mouth of the Río Negro and the mouth of the Río Aquidabán, and along the middle section of the Río Apa. The collecting efforts were directed towards an assessment of the decapod crustacean fauna, but other groups were incidentally collected. Sampled habitats included beaches of river, backwaters, lagoons, rapids, flooded plains, floating vegetation, submerged litter, and cryptic habitats. Collections were qualitative only and performed with seine nets, dip nets, traps, and by hand.

Thirteen species of decapod crustaceans were found: six species of Palaemonidae and one of Sergestidae shrimps, and six species of Trichodactylidae crabs. The occurrence of *Pseudopalaemon* sp. is the first record of this genus in the Río Paraguay basin, and is probably a new species. The shrimp *Macrobrachium amazonicum* and the crabs *Trichodactylus borellianus* and *Sylviocarcinus australis* were the most frequent species. The aquatic insect fauna was the richest in number of species, with 83 species in 34 families and ten orders. Coleoptera, Hemiptera, and Odonata were most often collected. Mollusks were an important group with regard to abundance and biomass, with at least seven species of Gastropoda and three species of Bivalvia. Other less representative groups included periphytic and parasitic crustaceans (Conchostraca, Ostracoda, Amphipoda, Isopoda, and Branchiura), annelids and platyhelminth worms.

Fishes

Two teams of ichthyologists sampled fishes at a combined total of 111 localities (Map 2). During the AquaRAP expedition, 136 fish species were collected in the Río Paraguay, 85 in the Río Apa, and 35 in the Riacho La Paz, for a grand total of 173 species.

Characiformes made up 54% of the fish fauna in the Río Paraguay. They belong to a variety of trophic categories, including herbivores, insectivores, mud-feeders, piscivores, and omnivores. Siluriformes (catfishes) and Perciformes (primarily cichlids) were the next most abundant groups. Piranhas (*Serrasalmus* spp.) were common and many fish had piranha bites. Some fishes (e.g., pacus, anostomids) feed on fruits and seeds, and depend on floods. Flooded forests are often used as nursery areas by many species of fishes, including commercially important species. Many taxa (e.g., *Lepidosiren paradoxa, Synbranchus marmoratus,* Callichthyidae) possess adaptations for breathing air. These are important for survival through the dry season. Only the eggs survive through the dry season in a number of species known as annual fishes (e.g., *Rivulus* sp.).

The diversity of fishes was lower in the Río Apa than in the Río Paraguay, probably because of the lower habitat heterogeneity. Characiformes, Siluriformes, and Perciformes were still the dominant taxa. Many species collected in the Río Apa were not collected in the Río Paraguay, highlighting the distinctiveness of the two basins. Several species of Siluriformes were found amidst the cover of logjams. Schools of *Salminus, Brycon,* and *Prochilodus* were observed migrating upstream.

Geographic and Habitat Partitioning of Fishes

The 173 species of freshwater fishes collected during the AquaRAP expedition to the Río Paraguay were analyzed to determine if any distributional patterns existed within the region collected. The region was divided into five sub-regions and also into 10 macro-habitats. Two null hypotheses were tested and rejected: (i) that fishes are randomly distributed with respect to sub-region; and (ii) that fishes are randomly distributed with respect to macro-habitat. The results showed that a strong sub-regional effect was evident such that two distributional elements were present consisting of: (i) an association between the Río Paraguay and Río Negro sub-regions and (ii) Río Apa and Riacho La Paz sub-regions. The analysis of distributions with respect to macro-habitats also reveals two components. The first relates the beach and main channel faunas to the macro-habitats inundated during seasonal flooding, such as flooded forests and lagoons. The assemblage associated with the flooding cycle accounts for more than 75% of the fishes collected. The second component, comprised of habitats found within the Río Apa and the Riacho La Paz (e.g., clear water, rapids, etc.), has a relatively sharp boundary with respect to the Río Paraguay showing more than a 50% turnover in fauna. These results show that core conservation areas must be designated to include areas were the seasonal cycle of flooding is unimpeded and the area of inundation is relatively unmodified. In addition, the Río Apa and Riacho La Paz represent a highly threatened area because of high rates of land conver-

sion, aridity, and the fact that much of the fauna cannot be recolonized from nearby water sources.

Congruence of Diversity Patterns among Fishes, Invertebrates, and Aquatic Plants

Null hypotheses concerning random distributions of species with respect to sub-regions and macro-habitats within the Río Paraguay are tested with data from 131 species of macro-crustaceans and benthic invertebrates and 186 species of aquatic plants. The patterns are compared to the results for the distributions of fishes presented by Chernoff et al. (2001). The invertebrate data demonstrate the identical pattern among sub-regions as evident in the fish distributions. The results support the recognition of two zones: (i) the Río Paraguay zone containing the Upper and Lower Río Paraguay and the Río Negro; and (ii) the Río Apa zone containing the Río Apa and the Riacho La Paz. For all data sets, the Río Paraguay zone has higher species richness than the Río Apa zone. The boundary between the two zones is abrupt, which is also supported by the plant data. Only 11 of 186 species of plants were found in both zones. There is no congruence of pattern among macro-habitats. Both invertebrate and plant data sets contain many values that are not different from mean random similarities. The fish data set provides an unambiguous pattern of habitats within the Río Paraguay zone tied together in relation to inundation of habitats during the flood cycle. The invertebrate set provides one cluster that is consistent with the general principal. The plant data set demonstrates a relationship among shore and sand habitats that experience greater currents than do other macro-habitats. The plants found in backwater habitats had little similarity to other macro-habitats. Based on these observations, it is concluded that significant habitat within each of the two zones needs to be preserved to maintain a large portion of the biodiversity.

Recommended Conservation and Research Activities

The basis for recommended conservation and research initiatives can be found in the executive summary and the chapters of this report. Many require international cooperation because the drainage systems originate in and flow through several countries, with some organisms (e.g., migratory fishes) routinely crossing national boundaries. Recommendations are listed in order of priority.

* Conserve portions of both the Río Paraguay **and** the Río Apa, not just one or the other. Emphasis should be placed on the Upper Río Paraguay-Río Negro and the Río Apa-Riacho La Paz, because together these two areas represent a significant proportion of the diversity within the region. The area surveyed by the AquaRAP expedition is particularly attractive for the establishment of preserves because

there are fewer people there than in the more heavily populated, southern Paraguay. Cooperation between Paraguay, Brasil, and Bolivia is encouraged since these drainages cross country boundaries. Specific attention should also be given to the Río Apa above Puerto San Carlos because tree species diversity is high, there have been few modifications to the ecosystem, and the soil is sandy and not fit for cattle ranching.

- Prevent large- and small-scale disruptions of the seasonal flooding. Interference or serious reduction of the flood cycle and inundated habitats will result in loss of wetlands (i.e., habitats for aquatic plants and invertebrates, nursery areas for fishes, etc.) and may result in up to a 50% loss of aquatic plant and fish species in the basin. Damming, dredging, straightening channels, and clearing rock outcroppings could all influence flooding.

- Study the impact of agricultural farms and cattle ranches on the environment. Determine levels of sustainable yield. Draft land-use legislation regulating slash-and-burn agriculture, logging, and human settlement. Prevent the logging of tree species in danger of being over-harvested.

- Conserve riparian vegetation. This will help to prevent erosion and maintain a source of nutrients for the aquatic systems. Reforestation programs should be implemented in riparian areas already damaged. Small streams, lagoons, flooded forests, and backwaters should not be overlooked as they demonstrate the highest vegetative diversity.

- Limestone areas should be conserved for their high water quality, and the Río Verde is under severe threat of salinization.

- There is no congruence among the fishes, invertebrates, and aquatic plants with respect to their distributions between macro-habitats. As a result, samples of all macro-habitats must be preserved to maintain the majority of species.

- Assess the reproductive and growth rates of commercial species of fishes, especially dorado, *Salminus maxillosus.* Determine levels of sustainable yield and adjust legislation accordingly. Monitor catches of fishes. International cooperation is required because fishes cross country borders.

- In its proposed state, the Hidrovia Project poses a major threat to the aquatic biodiversity of the Río Paraguay basin. The reduction of the number of fishes in the region will have important socio-economic consequences (e.g., loss or reduction of fisheries) in addition to the intrinsic value of biodiversity. If the Hidrovia Project is implemented, plans should be made ahead of time to

attempt to find adequate refugia within which the majority of the biodiversity can be self-supporting. For fishes, the Río Paraguay beach macro-habitat acts as a source for other macro-habitats from which recolonization could be possible. Some organisms, such as crabs and shrimps, are relatively easy to breed in captivity. Most species can be collected and maintained as "recolonization stocks" for a later date. This will help to preserve the uniqueness of the Paraguayan populations.

- Continue surveying the flora and fauna of the region. Ecological and systematic studies would be particularly valuable. Hydrological studies are also needed.

- Aquarium fishes (e.g., cichlids, tetras [characids], armored catfish [loricariids]) were collected during the expedition, particularly along Río Apa beaches and clear water macro-habitats. Unrestricted large-scale harvesting of these species would destroy their populations, as has occurred in some other areas of South America. After determining population sizes and conducting impact studies, it may be possible for local fishermen to implement a sustainable-yield fishery for aquarium fishes.

Literature Cited

Bonetto, A.A. 1986a. The Paraná River system. *In* Davies, B.R. and K.F. Walker (eds.). The ecology of river systems. Dordrecht, Netherlands: Dr. W. Junk Publishers. Pp. 541–555.

Bonetto, A.A. 1986b. Fish of the Paraná system. *In* Davies, B.R. and K.F. Walker (eds.). The ecology of river systems. Dordrecht, Netherlands: Dr. W. Junk Publishers. Pp. 573–588.

Chernoff, B., P.W. Willink, M. Toledo-Piza, J. Sarmiento, M. Medina, and D. Mandelburger. 2001. Testing hypotheses of geographic and habitat partitioning of fishes in the Río Paraguay, Paraguay. *In* Chernoff, B., P. W. Willink, and J.R. Montambault, (eds.). A biological assessment of the aquatic ecosystems of the Río Paraguay basin, Departamento Alto Paraguay, Paraguay. RAP Bulletin of Biological Assessment No. 19. Washington, DC: Conservation International. Pp. 80–98.

Ferreira, C.J.A., B.M.A. Soriano, S. Galdino, and S.K. Hamilton. 1994. Anthropogenic factors affecting waters of the Pantanal wetland and associated rivers in the upper Paraguay basin of Brazil. *In* Barbosa, F.S.R. (ed.). Acta Limnologica Brasiliensia, volume V, workshop: Brazilian programme on conservation and management on inland waters. Fundacão Biodiversitas—Sociedade Brasileira de Limnologia, Belo Horizonte, Brasil. Pp. 135–148.

Gleick, P. H. 1998. The world's water. The Biennial Report on Freshwater Resources. Washington, DC: Island Press. 308pp.

Goulding, M. 1980. The fishes and the forest. Berkeley: University of California Press. 280pp.

Heckman, C.W. 1998. The Pantanal of Poconé: biota and ecology in the northern section of the world's largest pristine wetland. Boston: Kluwer Academic Publishers. 622pp.

Lowe-McConnell, R. H. 1975. Fish communities in tropical freshwaters: their distribution, ecology and evolution. New York: Longman. 337pp.

Lourival, R. F. F., Silva, C. J, Calheiros, D. F., Albuquerque, L. B., Bezerra, M. A. O., Boock, A., Borges, L.M.R., Boulhosa, R.L.P., Campos, Z., Catella, A. C., Damasceno, G. A. J., Hardoim, E.L., Hamilton, S.K., Machado, F.A., Mourão, G. M., Nascimento, F.L., Nougueira, F.M.B., Oliveira, M.D., Pott, A., Silva, M.C., Silva, V.P., Strussmann, C., Takeda, A. M. and Tomás, W. M. (1999). Os impactos da Hidrovia Paraguai-Paraná sobre a biodiversidade do Pantanal - Uma discussão multidisciplinar. In Anais do II Simpósio sobre recursos naturais e sócio-econômicos do Pantanal. Corumbá - MS, Brasil, 517–534 pp.

Lundberg, J.G., L.G. Marshall, J. Guerrero, B. Horton, M.C.S.L. Malabarba, and F. Wesselingh. 1998. The stage for neotropical fish diversification: a history of tropical South American rivers. In Malabarba, L.R., R.R. Reis, R.P. Vari, Z.M.S. Lucena, and C.A.S. Lucena (eds.). Phylogeny and classification of neotropical fishes. Porto Alegre, Brasil: EDIPUCRS. Pp.13–48.

Moore, D. 1997. Hidrovia studies called profoundly flawed. World Rivers Review 12. 3pp.

Nakatani, K., G. Baumgartner, and M. Cavichioli. 1997. Ecologia de ovos e larvas de peixes. In Vazzoler, A. E. A. M., A. A. Agostinho, and N. S. Hahn (eds.). A planície de inundação do alto Rio Paraná: aspectos físicos, biológicos e socioeconómicos. Brasil: Editora da Universidade Estadual de Maringa. Pp. 281–306.

Pearson, N.E. 1937. The fishes of the Beni-Mamoré and Paraguay basins, and a discussion of the origin of the Paraguayan fauna. Proceedings of the California Academy of Science, 23:99–114.

Roberts, T.R. 1972. Ecology of fishes in the Amazon and Congo basins. Bulletin of the Museum of Comparative Zoology, 143:117–147.

Sowls, Lyle. 1997. Javelinas and other peccaries: their biology, management, and use. 2nd Ed. Arizona: Texas A & M University Press.

Switkes, G. 1996a. Belated public access to Hidrovia studies allowed. World Rivers Review, 10. 1pp.

Switkes, G. 1996b. Design chosen for first phase of Hidrovia. World Rivers Review, 11. 1pp.

Switkes, G. 1996c. Hidrovia plans propose heavy engineering for world's largest wetlands. World Rivers Review, 11. 2pp.

Switkes, G. 1997. Hydrologists condemn Hidrovia environmental studies. World Rivers Review, 12. 1pp.

Weil, T.E., J.K. Black, H.I. Blutstein, D.S. McMorris, F.P. Munson, and C. Townsend. 1984. Paraguay: a country study. Superintendent of Documents, Washington DC: U.S. Government Printing Office. 318pp.

WWF 2000. Retrato da Navegação no Alto Rio Paraguai. Relatório da Épedição Técnica Realizada entre os dias 3 e 14 de novembro de 1999 no Rio Paraguai entre Cáceres (MT) e Porto Murtinho (MS). Washington DC: World Wildlife Fund.

Resumen Ejecutivo (Español)

Introducción

El Río Paraguay es un río grande y con muchos meandros que, cuando se une con el Río Paraná aguas abajo, se forma la segunda cuenca más grande de América del Sur (Mapa 1). El tamaño de la cuenca oculta el hecho de que el Río Paraguay es relativamente muy poco profundo, descargando su porción de sólo 314 km³ de agua anual para la cuenca Paraguay-Paraná completa. Esta cantidad de agua es baja comparada con la cuenca del río Orinoco que descarga más de cuatro veces este valor (Gleick, 1998). El Río Paraguay, el cual tiene su cabecera en el planalto del estado de Mato Grosso en Brasil, corre por el humedal del Pantanal alrededor de la frontera oriental del Chocó y más abajo dividiendo al Paraguay en dos (Mapa 1). El Río Paraguay es un recurso acuático muy importante para la población de Paraguay, no sólo por su riqueza en biodiversidad, sino también por la dependencia humana en los recursos de pesquería, el agua dulce para el consumo humano y el riego para la agricultura.

La biodiversidad del Río Paraguay representa una transición interesante entre los ríos tropicales amazónicos al norte y los hábitats templados del sur. Las perturbaciones geológicas que ocurrieron hace 90 millones de años, han creado oportunidades para que tanto los ríos como las poblaciones existentes en sus cercanías cambiaran de cuencas entre los Ríos Amazona y Paraguay (Heckman, 1998: 31–32; Lundburg et al., 1998). Se supone que la similaridad entre la flora y fauna acuática existente en ambos pótamos es debido a intercambios del agua en las cabeceras de los ríos Guaporé y Arinos (los dos son de la cuenca amazónica) y las cabeceras del Río Paraguay durante los períodos de inundación (Pearson, 1937; Roberts, 1972; Lowe-McConnell, 1975: 47,85; Barnetto, 1986 a,b). Conexiones acuáticas del pasado y del presente nos ayudan a explicar la presencia de especies en Paraguay que se encuentra con frecuencia en la costa

atlántica o al sur de Argentina y Uruguay. Inclusivo los endemismos, la cuenca del Río Paraguay es un hábitat con asamblea de flora y fauna muy particular y único.

La caracterización de la flora y fauna del Río Paraguay es muy importante para documentar la diversidad acuática y el estado del ecosistema acuático y así poder determinar estrategias de conservación y desarrollo sostenibles para la región. Las amenazas pueden tomar muchas formas, pero la principal y más preocupante es el Proyecto planificado Hidrovía Paraguay-Paraná. La Hidrovía ("Hidrovía" quiere decir vía de agua), es un proyecto diseñado para canalizar 3,400 kilómetros de agua entre dos ríos: Paraguay y Paraná y poder hacerlos navegables por barcazas grandes y barcos de gran calado, para lo cual el dragado constante, la rectificación de meandros y la desaparición de los lechos rocosos son parte del Proyecto. El área afectada se extenderá desde Cáceres, Brasil en el Río Paraguay, hasta la desembocadura del Río Paraná en el océano Atlántico (Mapa 1). Muchas de las modificaciones realizadas en el río estarán en el Pantanal (Switkes, 1996 a,b,c), con la intención de facilitar el transporte principalmente de soja y hierro todo el año. Los países integrantes del Proyecto son: Argentina, Uruguay, Paraguay, Bolivia, y Brasil (Moore, 1997; Heckman, 1998: 497).

Las consecuencias no previstas en el Proyecto podría incluir: el desecamiento de grandes áreas del Pantanal y la bajada del nivel de las inundaciones en la parte superior del Río Paraguay y su consecuente aumento la parte inferior del río (Lourival et al., 1999; Switkes, 1996 b,c, 1997; WWF, 2001). La salinización de algunos tributarios del Río Paraguay, la contaminación del agua potable y un mayor riesgo de derrames de petróleo y sus derivados, causados por un mayor tráfico en el río, son también consecuencias probables de este Proyecto. Se cree que los beneficios económicos han sido exagerados debido a que no se han tomado en cuenta el transporte por carreteras y el potencial de competición con los ferrocarriles regionales (Switkes, 1996c, 1997; Moore, 1997; Heckman, 1998: 497–498).

La destrucción y conversión de hábitats no está restringido al proyecto Hidrovía. En efecto, los niveles de población están subiendo en Paraguay, conjuntamente con la agricultura, la explotación forestal y la ganadería. Estas actividades usualmente son más intensas en las cercanías de las comunidades, las carreteras y los ríos. La agricultura y la ganadería involucran además procesos de quema y deforestación de franjas grandes de vegetación natural. Cuando este proceso es constante, a largo plazo esta modificación del hábitat puede poner en peligro la biodiversidad además de aumentar los procesos de erosión.

La Expedición AquaRAP

Para tratar de evaluar las consecuencias del Proyecto Hidrovía y otras amenazas, un equipo multidisciplinario e internacional se hizo cargo de la expedición de Evaluación Rápida de Ecosistemas Acuáticos (AquaRAP) en la cuenca del río Alto Paraguay con los objetivos de: (i) describir los aspectos biológicos y ambientales previos a las modificaciones a gran escala y (ii) determinar, si fuera posible, los cambios potenciales en los recursos acuáticos debidos a las amenazas. La expedición se llevó a cabo entre el 4 y el 18 de septiembre de 1997 y se incluyó la sección del Río Paraguay entre el Río Negro, al norte y el río Aquidabán, al sur (Mapa 1). Porciones del Río Apa y el tributario riacho La Paz también fueron investigados. Para realizar mejor las comparaciones a lo largo del río, el área de la expedición fue dividida en cinco regiones: Río Negro, río Alto Paraguay (la parte superior del río desde el cerro Pão de Açúcar al norte), Río Apa y riacho La Paz (Mapa 1). Fueron evaluadas: la vegetación terrestre, la vegetación acuática, las características del agua, tanto físicas como químicas, el plancton, los macro-invertebrados acuáticos y los peces. Los resultados de estas evaluaciones pueden ser tomadas en cuenta como base para las propuestas de recomendaciones respecto del Proyecto Hidrovía y otras amenazas, así como para la planificación de futuras actividades de investigación y conservación en el área.

El equipo científico fue internacional y multidisciplinario. Participaron equipos de Bolivia, Brasil, Perú, Paraguay, Venezuela y los Estados Unidos. Los científicos fueron especialistas en el área de botánica (vegetación terrestre y acuática), crustáceos, macro-invertebrados, genética de poblaciones, ictiología y limnología, incluyendo química del agua y plancton. El equipo AquaRAP se conformó tanto por científicos como por estudiantes locales. Si bien la expedición enfocó especialmente el valor biológico y de conservación de la región, se contemplaron también las amenazas actuales y futuras a las que se enfrentará esta región. La expedición además evaluó la metodología y los protocolos empleados para una evaluación rápida en los ecosistemas acuáticos.

El Programa de Evaluación Rápida de Ecosistemas Acuáticos de Sudamérica es un programa internacional y multidisciplinario cuyo objetivo es el de determinar áreas prioritarias de conservación y de uso sostenible en ecosistemas de agua dulce en América Latina. Su función es evaluar el valor biológico y la necesidad de conservación de ecosistemas acuáticos tropicales por medio de inventarios rápidos y hacer llegar los resultados sin demora a los responsables sean políticos, científicos, conservacionistas y agencias internacionales. AquaRAP es un programa de colaboración entre Conservación Internacional (CI) y the Field Museum.

La estructura de AquaRAP es la de un Comité conformado por un equipo internacional de científicos de siete países (Bolivia, Brasil, Ecuador, Paraguay, Perú, Venezuela, y los Estados Unidos). El Comité revisa y controla el protocolo y los parámetros utilizados para la selección de las áreas prioritarias de conservación para las evaluaciones rápidas. Las expediciones, las cuales implican una mayor colaboración entre los científicos de América Latina, también promueven un intercambio internacional y oportunidades de capacitación. La información recogida en las expediciones de AquaRAP es repartida por medio de una serie de Publicaciones de Evaluación Rápida realizada por Conservación Internacional en Washington y que fuera diseñada para los responsables de las decisiones locales, los políticos, los líderes y los conservacionistas en general, quienes pueden imponer prioridades de conservación y guiar las acciones a través de los fondos que llegan a la región.

Una evaluación de la biodiversidad en los ríos Paraguay y Apa fueron muy importantes antes de hacerse modificaciones a gran escala en la zona. Esta parte de Paraguay lastimosamente, es muy poco conocida por la ciencia. La mayor parte de las colecciones en el pasado, fueron hechas en la cercanía de los centros de población, con énfasis en la región Oriental, sobre la margen izquierda del Río Paraguay. Informaciones adicionales fueron necesarias para entender mejor las oportunidades de manejo sostenible. La sección que sigue da un resumen de las observaciones realizadas durante la expedición AquaRAP; una información más detallada se tiene en cada uno de los capítulos; los resúmenes de los mismos se encuentran después de la síntesis mencionada anteriormente, seguidos de las recomendaciones de las actividades de investigación y conservación.

Resumen

La expedición AquaRap ayudó a aumentar los conocimientos sobre la región, debido a que antes de que se realizara dicha expedición los conocimientos sobre la misma se limitaban únicamente a inventarios forestales, estudios hidrológicos y algunos estudios sobre la salinidad de los suelos. Esta expedición recopiló datos actualizados sobre botánica, limología, ictiología e invertebrados. El poder integrar esta nueva información nos permite hacer un enfoque más general y comprensivo del ecosistema acuático en su totalidad. Basado en este enfoque general, hacemos las observaciones y recomendaciones en la sección siguiente.

Río Paraguay

La sección más nórdica del Río Paraguay en Paraguay, está ubicada entre el Pantanal y el Chaco, que son dos tipos de hábitats que no se pueden encontrar en ninguna otra parte del planeta. El Pantanal es el humedal más grande del mundo y es reconocido por sus bandadas de pájaros, sus numerosos caimanes, carpinchos (capibaras), ciervos, osos hormigueros y otras formas de animales. La planicie abierta hace que éste sea un paraíso para poder observar a dichos animales. El Chaco es una enorme sabana de belleza sin igual, cubierta de pastizales, bosques xeromorfos y matorrales. El agua es escasa y solamente se reabastece durante el verano y el periodo de inundaciones estaciónales. La flora y fauna del Chaco incluyen a: jaguares, pumas, ñandúes, ciervos, manadas de zorros, anacondas y otro tipo de serpientes como la yarará y la cascabel, osos hormigueros, tapires, carpinchos pecaríes, incluyendo el taguá (*Catagonus wagneri*—al que se creía ya extinto) (Sowls, 1997), y numerosas bandadas de aves, muchas migratorias y semejantes a las del Pantanal, los que se han podido adaptar a las duras condiciones del terreno y sorprendentemente son muy numerosas. La cantidad y diversidad de estos organismos es mayor a lo largo de los ríos y pantanos (Weil et al., 1984: 17–19). La originalidad de ambos ecosistemas a nivel global (Pantanal Chaco) y su dependencia con los fenómenos hidrológicos, hacen que estos sean muy favorables para las iniciativas de conservación en ambientes acuáticos a condición de que estas regiones no hayan sido severamente dañadas por actividades las humanas.

El área que fuera estudiada por la expedición AquaRAP se encuentra en buen estado. Los habitantes de la región dependen muchísimo del agua para poder vivir; el área presenta bajo número de pobladores, agricultores y ganaderos. La productividad es alta en cuanto a las macrófitas acuáticas y ellas son las responsables de la descomposición de la materia orgánica y posterior transformación y producción de nutrientes para las aguas del río. Los recursos de pesquería son muy poco utilizados por la población. No se encontró contaminación directa en el agua. Desafortunadamente esta situación probablemente pueda cambiar.

Se espera que el número de habitantes crezca. Las posibilidades de desarrollo en la región y el crecimiento descontrolado pueden causar problemas innecesarios a largo plazo. Es sumamente importante que se realicen estudios sobre el impacto que pueda ocasionar el mantener la producción ganadera y pesquera a niveles sostenibles. En efecto, el aumento de la agricultura y la ganadería aumentan la erosión del suelo, con la cual se pierden especies ictícolas muy valiosas a nivel comercial, así como ya está ocurriendo cerca de Bahía Negra. La contaminación de las aguas en el futuro, a través de la industria y los habitantes, pueden causar

eutrofización y problemas de salud a sus residentes. Es sumamente importante como medida preventiva el mantener la vegetación ribereña, con el objeto de limitar la erosión y así mantener menos turbias las aguas y también tratar a las aguas negras. Hay posibilidades de que se ejerza presión sobre cantidad de pesca comercial y turística, las cuales tienen un impacto en la reproducción y crecimiento de los peces y es necesario que se tenga en consideración este punto cuando se estén escribiendo nuevas leyes. Si se ignora la dinámica de la población ictícola, solo es cuestión de tiempo para que la pesca sufra un colapso. En general es necesario que se realicen en la región más trabajo de investigación en el área ecológica para así poder determinar con más exactitud los niveles de sostenimiento agrícolas versus recursos pesqueros.

Quizás el peligro más grande a corto plazo es el Proyecto Hidrovía Paraguay-Paraná, el cual habilitará a embarcaciones de gran calado y furgones de barcazas a navegar por el Río Paraguay. Todavía no se sabe cual sería el impacto exacto que este Proyecto podría causar a dicha región. (ver Mapa 1) Sin embargo se anticipa que algunas partes del Pantanal en las cercanías del Río Paraguay se secarán y aumentarán las inundaciones en las áreas más bajas del río (Lourival et al., 1999; Switkes, 1996b,c, 1997; WWF, 2001). Es muy difícil poder predecir el impacto hidrológico que habrá en el área en donde se realizó el estudio pero de cualquier manera las consecuencias alterarán la distribución actual y la composición de las comunidades acuáticas. La realidad es que es muy difícil predecir el daño que tendrá en su totalidad el dragar y canalizar el río, pero esto es ya suficiente para estar preocupados. La reducción en las inundaciones y como resultado, la pérdida de las áreas de humedales sería devastador para las comunidades acuáticas. El fenómeno ecológico dominante en la región son las inundaciones estaciónales, con organismos locales acuáticos y semi-acuáticos que se adaptan a los períodos de inundación. Estas inundaciones redistribuyen los nutrientes y facilita el crecimiento de la vegetación de los humedales. Muchas plantas dependen de las inundaciones para dispersar sus semillas (Goulding, 1980). El ascenso de los niveles del agua también sirven como señal para la reproducción de muchas especies comerciales de peces de gran valor, en los que para muchos de los organismos, el mayor crecimiento ocurre durante la época de inundaciones (Lowe-McConnell, 1975; Goulding, 1980; Bonetto, 1986b). Las áreas inundadas actúan como refugio para muchas especies de peces e invertebrados. La interrupción de las inundaciones disminuirá dramáticamente la heterogeneidad del hábitats, la biodiversidad y la biomasa.

Los efectos de la Hidrovía Paraguay-Paraná no se limitarán solamente al ecosistema acuático; en efecto, el número de individuos de peces se verá reducido y esto sería un desastre para la población que vive de ello. Aves, mamíferos y reptiles que cuentan con macroinvertebrados acuáticos y peces para su alimentación, también se verán afectados. Las inundaciones aguas abajo del Río Paraguay destruirá las tierras aptas para cultivo y forzará a la población que vive cerca del río a tener que migrar.

Río Apa

El Río Apa se presenta en buenas condiciones. La calidad del agua es alta y los niveles de población todavía son bajos. La actividad de los habitantes de la región es la agricultura, la ganadería y la explotación forestal. El río provee agua limpia y comida para las comunidades locales y varias aves, reptiles, y mamíferos.

Aunque la parte baja del Río Apa, cercana a su desembocadura en el Río Paraguay es similar a este último, la parte alta del río es ecológicamente diferente, con barrancos empinados con afloramientos de bancos arenosos. Una flora y fauna diversa reflejan la diferencia del hábitat entre ambos ríos. Deberían hacerse esfuerzos para que ciertas áreas de la parte baja y parte alta del río sean conservados. La heterogeneidad de la región así como su biodiversidad podrán ser mantenidas únicamente si se pueden conservar ciertas áreas de ambos ríos.

Con sus barrancos empinados y suelos no consolidados, el Río Apa es muy susceptible a la erosión. La vegetación ribereña ha ayudado a mantener el suelo en buen estado. Esta manera natural y efectiva de mantener los suelos está amenazada por la explotación forestal en la que la tala de los árboles y las quemas para la preparación de los suelos agrícolas, remueven la vegetación natural necesaria para la protección de los suelos.

La erosión puede causar muchos problemas. Uno de ellos es la pérdida de la capa superior del suelo lo cual eventualmente deja un suelo menos apto para la producción de la cosecha. Esto tendrá un impacto en la agricultura la cual solamente se podrá sostener por medio del uso de fertilizantes. Estos finalmente serán arrastrados al río, pudiendo a largo plazo causar eutrofización en algunas áreas del río. Otro problema más grave y de menor plazo es que el nivel de sedimentos en el río puede subir, lo que aumentaría la erosión y hará que las aguas del río se pongan turbias, lo cual afectará este hábitat muy particular y único y consecuentemente se verá reducida la heterogeneidad regional, tanto del hábitat como de la biodiversidad.

Una de las primeras razones por la cual la tala, quema y la explotación de los bosques pone en peligro al área es por la falta de buenas y estrictas regulaciones o leyes. Si no se toman medidas para limitar y controlar la destrucción del hábitat, especialmente las áreas de la ribera del río, se producirían problemas a largo plazo al aumentar los niveles del sedimento y la erosión. Reglas o leyes que aseguren o indiquen cuáles tipos de árboles se pueden cortar asegurarán a largo plazo a la región éxitos económicos y se podrán prevenir la extinción de valiosas maderas como es la de *Amburana caerencis,* el "trébol". Con un plan de mantenimiento adecuado existe la posibilidad de mantener la productividad

mientras se mantenga la integridad biológica del ecosistema acuático.

El Proyecto Hidrovía tendrá un impacto en la parte baja del Río Apa casi de la misma manera como lo planteado para el Río Paraguay. Las consecuencias hidrológicas en la parte alta del Río Apa son poco claras debido a que el río se encuentra más elevado que el Río Paraguay. Existe la posibilidad de que las aguas subterráneas drenen con más rapidez reduciendo el tamaño del río. Dentro de este escenario, las poblaciones de organismos acuáticos se verían reducidas pero los tipos de hábitats y la diversidad no sufrirían cambios a menos que el volumen de la corriente del río baje sustancialmente. No se anticipan inundaciones en la parte alta del Río Apa.

Resumen de Capítulos

Descripción Geográfica

El Río Paraguay tiene una extensión de 2,550 kilómetros (1,580 millas) y su cuenca llega a cubrir un área de 1,100,000 kilómetros cuadrados (Bonetto, 1986a). El río comienza en la parte del Planalto de Mato Grosso denominado Chapada dos Parecis en la parte sudoeste de Brasil y subsecuentemente navega hacia el sur a lo largo del Pantanal, el humedal más grande del mundo (mapa 1). Luego de salir del Pantanal, el Río Paraguay se transforma en límite con el Brasil, hacia el este y el Chaco paraguayo al oeste y luego de la desembocadura del Río Apa, el Río Paraguay divide al país (Paraguay) en dos regiones naturales muy distintas: la Occidental o Chaco y la Oriental. Mas al sur, el Río Paraguay se une al Río Paraná y juntos forman la segunda cuenca más grande de América del Sur.

A pesar de que el nivel de los ríos depende de la cantidad de lluvias, la relación no es tan simple. En el Pantanal las lluvias ocurren durante los meses de verano, pero las pendientes bajas y las inundaciones de los humedales hace que el agua se mantenga en esta parte por varios meses. Esta agua lentamente va corriendo hacia el sur. De acuerdo con Ferreira et al. (1994) el promedio de los niveles más altos de las aguas de la cuenca del Río Paraguay ocurren de la siguiente manera: en la parte noreste del Pantanal desde febrero a marzo; en Corumbá desde abril a mayo y en la desembocadura del Río Apa desde julio a agosto. Este fenómeno hidrológico altera la magnitud de la corriente del río aguas abajo. Aunque el nivel de las aguas en el Pantanal al norte puede llegar a cambiar hasta 4 metros (13 pies), el Río Paraguay cerca de Asunción solamente llega a elevarse 1.5 metros (5 pies), (Weil et at., 1984:15)

En la parte norte del Paraguay, el Río Paraguay es como una enorme serpiente que va atravesando terrenos relativamente planos. El nivel de fluctuación de las aguas conjuntamente con las planicies hacen que se extiendan las inundaciones, las cuales afectan el curso del río. Es muy común encontrar lagunas y canales a través del río así como las conexiones de los bancos arenosos con los bosques y los palmares. Las formaciones mencionadas son rutinariamente inundadas durante varios meses del año, dependiendo esto de la altura de las mismas. Masas enormes de vegetación flotante (por ejemplo: *Eichhornia* sp., "camalotes"), son tan grandes que se conocen mejor como "praderas flotantes", o "embalsados". El afloramiento de rocas en el área de los cerros (como: Pão de Açúcar), forman un embotellamiento natural que retrasa el flujo del agua río abajo.

La diferencia entre el Río Apa y el Río Paraguay es que es mas pequeño y presenta cascadas (rápidos); el sedimento predominante en el fondo es la arena y sus barrancos son más altos y están bien definidos. Es muy común ver la acumulación de troncos y ramas de los árboles en el canal del río. La diversidad del hábitat en el Río Apa es menor a la del Río Paraguay y el nivel de las aguas puede cambiar muy rápidamente, dependiendo esto de las lluvias.

El riacho La Paz se puede describir como la de un río mas pequeño con un curso cambiante entre rizos y pozos, con abundantes meandros. El fondo de este es rocoso y el sedimento predominante es la arena; el nivel de las aguas del río es similar a las del Río Apa es decir, cambia con rapidez dependiendo de las lluvias. En general presenta una diversidad de hábitat bastante baja.

La Vegetación Terrestre

12 variables transectos fueron establecidos para el estudio de la vegetación terrestre a lo largo del Alto Río Paraguay y Río Apa. Las colecciones generales fueron hachas también en otras partes fuera de los transectos. Un total de 198 especies de plantas fueron identificadas. Se encontraron 5 tipos básicos de hábitats a lo largo del Río Paraguay: de inundación estacionarais, manchones de palmares de *Copernicia alba*, manchones de palmares no inundados de *Copernicia alba*, bosques modificados de *Schinopsis balansae*, bosques sobre mesetas y el bosque xerófito chaqueño.

Los manchones de palmares de inundaciones estaciónales son muy comunes a lo largo del Alto Río Paraguay. La especie dominante es *Copernicia alba* pero además arbustos, lianas y herbazales son también abundantes. Los manchones no inundados de palmares de *Copernicia alba* son a menudo modificados y convertidos en pastos. El resultado es un modificado que contiene abundantes arbustos pioneros que modifican este tipo de formación como ser: *Acacia caven, Prosopis ruscifolia,* y *Mimosa pellita* entre otras. El bosque modificado de *Schinopsis balansae* estaba compuesto por arbustos como: *Coccoloba guaranitica, Ziziphus mistol, Prosopis ruscifolia, Caesalpinia paraguariensis, Copernicia alba* y *Pseudobombax* sp.

La altura de los árboles están en relación con las inundaciones y los suelos drenados rápidamente; es por esta razón que la flora establecida es similar con aquellas de pequeñas áreas inundadas. Los árboles tienden a ser pequeños y las hierbas efímeras. El bosque chaqueño es similar, con árboles alcanzando 9–12 metros de altura. El sotobosque se encuentra dominado por arbustos espinosos y herbazales.

Se han encontrado 3 tipos básicos de hábitats a lo largo del Río Apa: la vegetación ripiaría o marginal, el bosque semi-deciduo y el cerradón. La vegetación marginal se encuentra en los bordes del Río Apa y su área de inundación, durante el periodo de "aguas altas".

El bosque semi-deciduo puede estar dividido en 3 pisos de vegetación: el del piso más alto, compuesto por árboles de 25–30 metros de altura; el piso intermedio de árboles entre 15–20 metros de altura, tales como: *Croton urucurana, Trichilia* sp, *Astronium fraxinifolium* y el piso más bajo donde se encuentran algunos arbustos hasta 1 metro de altura o más, formando matas o arbustales poco densos (matorrales).

Muchas de las especies son deciduas y cuando se originan lluvias, los cerradones se asemejan a un bosque húmedo, debido a la densidad de árboles. Las formaciones de cerradones son las que se encuentran más lejanas del río, donde los grandes árboles son raros y poco frecuentes.

Por encima de todo eso, la vegetación terrestre adyacente a los ríos se encuentra muy influenciada por las fluctuaciones del agua (crecientes anuales) y tiempos lluviosos.Las modificaciones de los hábitats, debido principalmente a los incendios y a la ganadería, han tenido y tienen un impacto significativo en la región. Evidencias de la vegetación prístina no fueron encontradas en la región durante la expedición.

La deforestación excesiva acelera los procesos de erosión, resultando una mayor sedimentación en el Río Paraguay.

La Vegetación Acuática

La vegetación acuática fue muestreada en 31 localidades a lo largo del Alto Río Paraguay (Mapa 2). Un total de 147 especies fueron identificadas a lo largo del Río Paraguay y 50 especies a lo largo del Río Apa, haciendo un total de 197 especies de plantas relacionadas directamente con el agua.

A lo largo del Río Paraguay, la vegetación dominante es la de un "mosaico de vegetación" compuesto por: los bosques semi inundables de *Schinopsis balansae*, las sabanas hidromórficas de *Copernicia alba* (sabanas palmares) y la vegetación acuática ubicada en las depresiones con agua permanente (lagunas).

Los hábitats más comunes para la vegetación acuática han sido divididos en: la vegetación de ambientes semi-lóticos, vegetación de pantanos, lagunas, bancos arenosos y los embalsados. Las especies fueron categorizadas según su estilo de vida en: epifitas, lianas, flotantes libres, emergentes, sumergidas, enraizadas en el fango y la vegetación palustre de los embalsados y ambientes inundables temporariamente en general.

Los hábitats mixtos con corrientes muy lentas de agua (ambientes semi-lóticos), aparecen en donde el río penetra sobre los barrancos e inunda temporariamente los suelos;

estos hábitats resultaron ser mucho más diversos de lo esperado y probablemente se han constituido en las áreas más
diversas para la flora acuática dentro del área de estudio.

Esta alta diversidad podría estar relacionada por la
dinámica del agua y el ciclo de los nutrientes. Las depresiones inundadas temporariamente son relativamente
pobres en vegetación, posiblemente debido a que el constante movimiento de las aguas impida el crecimiento de
muchas especies.

Los embalsados son muy variables; constituyen grandes
masas de vegetación flotante, algunos con un sustrato o suelo
con sedimentos muy escasos que podrían ser mejor descrip-
tos como "pastizales de vegetación flotante". Otros desarrol-
lados usualmente en áreas más protegidas contra los vientos,
se encuentran en un estado más avanzado de desarrollo, con
un apreciable sustrato como suelo. El número de especies de
estos embalsados es proporcional al espesor del sustrato.

Los bancos arenosos tienen un número generalmente bajo
en número de especies, probablemente debido a la pobreza
de nutrientes en los suelos, lixiviados por la acción del agua.
Sin embargo, las especies encontradas en estos ambientes son
muy distintas y características, diferentes a los demás hábitats, de ahí su importancia.

El Río Apa presenta una diversidad mucho menor en su
flora acuática que el Río Paraguay; esto es debido probablemente a que la velocidad de subida y bajada de las aguas del
río es mucho mayor, lo que produce una carencia en la diversidad de hábitats con preponderancia de las arenas (afloramiento de pequeños bancos de arena) y un gran número
de troncos y ramas flotando sobre el canal del río.

Los hábitats más diversos fueron los encontrados en
algunos de los tributarios de este río.

Calidad del Agua y Macro-Invertebrados Bénticos
Se evalúo la calidad del agua de 35 localidades (Figura 4.1).
Las aguas del Río Paraguay son por lo general ligeramente
ácidas (pH 6.0–6.5) con un nivel de oxígeno bajo, (<6.0
mg/L), con baja conductividad eléctrica (60–100 µS/cm) y
con temperaturas que van de los 24 hasta los 27 centígrados.
Las aguas del Río Apa son neutras a poco alcalinas (pH
7.27–7.96), con niveles variables de oxígeno (4.2–7.13
mg/L), alta conductividad eléctrica (163-376 µS/cm), y temperaturas que oscilan entre los 22 a los 29 centígrados.

Las comunidades de fitoplancton eran relativamente
diversas e incluían representantes de varios grupos de ellas
incluyendo Chlorophyta, Euglenophyta, Chrysophyta,
Bacillariophyta, y Cyanophyta. Se recolectaron macro-invertebrados bénticos en 33 lugares. Chironomidae (Diptera) larvae y Oligochaeta fueron los grupos dominantes de 27 de los
lugares así representando el 52% y 35% de los organismos
registrados. Se recolectaron con menos frecuencia Odonata,
Trichoptera, Ephemeroptera, Chaoboridae, Ceratopogonidae, Corixidae, Conchostraca, Planaria, Nematoda,
Hirudinea, Copepoda, Ostracoda, Bivalvia, y Gastropoda.

Entre las Chironomidae larvae, se pudieron identificar 26
taxa, algunas pueden ser especies aún no descriptas. La diversidad de macro-invertebrados bénticos es comparable a la de
otras cuencas en Sudamérica. La mayoría de los géneros son
típicos de los humedales herbáceos, estanques, lagos y porciones del río que tienen movimiento lento. Las áreas ricas en
materia orgánica presentan una baja diversidad. En total hay
muy poca evidencia de que exista directa contaminación de
las aguas. La única excepción podría ser cerca del pueblo
(pequeña ciudad) de Bahía Negra donde la deforestación y la
quema podrían traer como resultado la erosión y
eutrofización, a largo plazo.

Macro-invertebrados
Se examinaron los macro-invertebrados de 54 estaciones a lo
largo de la parte alta del Río Paraguay entre la boca del Río
Negro y la boca del río Aquidabán, y a través de la mitad del
Río Apa. Los esfuerzos colectivos fueron dirigidos hacia la
evaluación de la fauna crustáceo decápodo pero otros grupos
fueron accidentalmente recogidos. Los tipos de hábitats
muestreados fueron los siguientes: playas, aguas estancadas,
lagunas, rápidos, terrenos aluviales, vegetación flotante
(embalsados), vegetación sumergida y otros hábitats no
definidos. La recolección fue cualitativa y se llevó a cabo solamente con redes de jábega, redes profundas, trampas y con
las manos.

Se encontraron 13 especies de crustáceos decápodos: seis
especies del camarón Palaemonidae y una especie de Seges-
tidae, y seis especies de cangrejos Trichodactylidae. Se encontró por primera vez en la cuenca del Río Paraguay la presencia de la especie *Pseudopalemon* sp. y probablemente sea una
nueva especie. Las especies que se encontraron con más frecuencia fueron la del camarón *Macrobrachium amazonicum*
y los cangrejos *Trichodactylus borellianus* y *Sylviocarnicus aus-
trales*. La fauna de insectos acuáticos fue la más numerosa con
83 especies de 34 familias y diez ordenes. Las especies más
colectadas fueron Coleoptera, Hemiptera y Odonata. Los
moluscos ocuparon un grupo importante con respecto al la
abundancia de biomasa con por lo menos siete especies de
Gastropoda y tres especies de bivalvia. Otros grupos que
fueron menos típicos eran los crustáceos periphyticos y para-
siticos (Conchostraca, Ostracoda, Amphipoda, Isopoda y
Branchiura), gusanos annelis y platyhelmith.

Peces
Dos grupos de ictiólogos tomaron muestras de los peces en
un total de 111 localidades (Mapa 2). Durante la expedición
de AquaRAP se recogieron del Río Paraguay 136 especies de
peces, 85 en el Río Apa y 35 en el riacho La Paz con una
suma total de 173 especies.

Los Characiformes constituyen el 54% de toda la fauna
de peces del Río Paraguay. Ellos pertenecen a una variedad de
categorías trópicas, incluyendo a los herbívoros, los insec-
tívoros, los que se alimentan del lodo, los que comen peces y

los omnívoros. La próxima categoría más abundante fue la de los Siluriforms (bagre) y Perciformes (primeramente cichlids). Fue muy común encontrar pirañas (*Serrasalmus* spp.) y además se encontraron muchos peces que presentaban mordidas de ellas. Alguno de los peces (por ejemplo pacus y anostomids) se alimentan de frutas y semillas dependiendo de las inundaciones. Las inundaciones forestales son casi siempre áreas utilizadas como criadero para muchas especies de peces incluyendo especies comerciales muy importantes. Muchas taxa (por ejemplo: *Lepidosiren paradoxa, Synbranchus marmoratus,* Callichthyidae) poseen adaptaciones para poder respirar aire; estas adaptaciones son muy importantes para poder sobrevivir durante los meses de sequía. Solamente los huevos sobreviven de los meses de sequía en muchas de las especies conocidas como peces anuales (por ejemplo: *Rivulus* sp.).

En el Río Apa se encontró una diversidad más baja que la del Río Paraguay, esto es debido a que probablemente existen una heterogeneidad más baja de hábitat. Las taxa dominantes fueron los Characiformes, Siluriformes y los Perciformes. Muchas de las especies que se recolectaron en el Río Apa no fueron recolectadas en el Río Paraguay, por lo que se destaca la diferencia entre ambas cuencas. Muchas especies de Siluriformes se encontraron en medio de los troncos flotantes. Se observaron varios cardúmenes de *Saminus, Brycon* y *Prochilodus,* los que migraban aguas arriba del río.

División de Peces por la Geografía y el Hábitat

Fueron analizas las 173 especies de peces de agua dulce que se recolectaron durante la expedición AquaRAP en el Río Paraguay para poder determinar si existe dentro de la región patrones de distribución. La región fue dividida en cinco sub-regiones y también en diez macro hábitats. Se examinaron y se rechazaron dos hipótesis nulas: (i) que los pescados se distribuyen en las sub-regiones al azar, y (ii) que los pescados se distribuyen al azar con respecto al macro hábitat. Los resultados mostraron que había un efecto sub regional debido a que dos elementos de distribución están presente los cuales consistían en (i) la asociación de las sub-regiones entre el Río Paraguay y Río Negro (ii) y las sub-regiones del Río Apa y riacho La Paz. Se encontraron dos componentes en los análisis de distribución con respecto al macro hábitat. El primer componente se relaciona con las playas y con los principales canales de fauna y macro hábitats que se inundan durante la época de crecientes así como en los bosques inundados y lagunas. Un 75% del total de especies fueron recolectados en el Río Paraguay, hecho que se relaciona con el periodo de inundaciones. El 50% de los peces encontrados en el Río Apa son diferentes a los del Río Paraguay, debido a sus condiciones físicas como aguas más claras, presencia de rápidos, etc.).

Estos resultados muestran que el foco de conservación para dichas áreas debe también aquellas hasta donde llegan los ciclos de las inundaciones. Adicionalmente el Río Apa y el riacho La paz representan una área altamente amenazada debido a que existe un alto nivel de modificación de las tierras para la agricultura; por otro lado, no se han podido realizar buenas colecciones en esta área.

Congruencia de patrones de diversidad entre los peces, invertebrados y plantas acuáticas

La hipótesis nula que se refiere a la distribución al azar de las especies con respecto a las sub-regiones y a los macro hábitats del Río Paraguay fueron examinados con datos de 131 especies de macro crustáceos e invertebrados bénticos y 186 especies de plantas acuáticas. Los patrones fueron comparados a los resultados de distribución de pescados presentados por Chernoff et al. (2001). Los datos de los invertebrados demostraron patrones idénticos a aquellos de las sub-regiones como evidencia en la distribución de pescados. Los resultados apoyan el reconocimiento de dos zonas: (i) la del Río Paraguay, zona que se extiende desde el Alto Paraguay hasta el sur, incluyendo el Río Negro; (ii) la del Río Apa, zona que contiene el Río Apa y el riacho La Paz. El conjunto de datos demuestra que para ambas zonas, la del Río Paraguay contiene más cantidad y riqueza de especies comparada con la del Río Apa. Los limites entre las dos zonas son muy diferentes y esto se comprueba por medio de los datos aportados por las plantas. Solamente once de 186 especies de plantas fueron encontradas en las dos zonas. No ha habido congruencia entre los patrones entre los macro hábitats. Ambos datos, los de invertebrados y los de plantas contienen muchos valores que no son diferentes a los del promedio de las similaridades al azar. Los datos de los peces proveen un patrón ambiguo en los hábitats dentro que corresponden a la zona del Río Paraguay.

Actividades Recomendadas para la Conservación e Investigación Científica

Las recomendaciones para las actividades de conservación e investigación puede ser encontradas en el resume ejecutivo y en los capítulos de este informe. Muchas de estas requieren la cooperación internacional porque las cuencas tienen su orígenes y vías también por otros países; otra razón es que los peces migratorios no tienen fronteras y se transladan de un país a otro. Se presentan aquí algunas recomendaciones, en orden de prioridad.

- Conservar porciones de ambos Ríos (Paraguay, Apa), no solamente una de ellas. Se debe poner énfasis en la zona del río Alto Paraguay-Río Negro **y** el Río Apa-riacho La Paz, porque estas dos zonas representan una porción significativa para la diversidad dentro la región. El área estudiada por la expedición AquaRAP es muy apropiada para hacer reservas porque hay menos población en esta área que aguas abajo. La cooperación entre Paraguay, Brasil y

Bolivia debe ser alentada porque estas cuencas cruzan sus fronteras. Atención especial se debe al Río Apa arriba de Puerto San Carlos porque la diversidad de especies de árboles es alta, el ecosistema de esta zona ha sido muy poco modificado, y los suelos son arenosos y así no son apropiados para la ganadería.

- Prevenir el nivel de las inundaciones temporales a diferentes escalas (mayor y menor). La reducción significativa de las inundaciones resultaría en la pérdida de los humedales, lo que equivale también a la pérdida de hábitats para las plantas, invertebrados acuáticos, áreas de criaderos de peces, etc.; finalmente esto podría resultar en una pérdida de casi el 50% de las especies de plantas aquáticas y peces en la cuenca. La construcción de presas, canales en el río, rectificación de meandros y pulverizando las rocas podrían afectar a las inundaciones naturales.

- Estudiar el impacto que tiene la agricultura y la ganadería en el medio ambiente. Determinar los niveles de cosecha sostenible. Escribir una legislación apropiada para el uso de los suelos y regular los métodos agrícolas que se valen de la deforestación y la quema para la preparación de los suelos. Prevenir el corte comercial de las especies arbóreas que están en peligro de extinción.

- Conservar la vegetación ribereña. Esto ayudaría a prevenir la erosión y mantener la fuente de nutrientes para los sistemas acuáticos. Programas de reforestación deben ser implementados en áreas ribereñas que ya se encuentran dañadas. Hábitats como ríos muy pequeños, lagunas, bosques inundables, y aguas estancadas deben ser tomados en cuenta porque representan la diversidad de vegetación más alta.

- Áreas de piedra caliza deben ser conservadas por su calidad de agua muy buena, y el Río Verde está muy amenazado de ponerse salinizado.

- La distribución de peces, invertebrados, y plantas acuáticas no es constante en los macrohábitats. Por lo tanto, se deben preservar muestras de todos los tipos de hábitats para mantener la mayoría de las especies existentes.

- Evaluar el nivel reproductivo y de crecimiento de las especies comerciales de los peces, especialmente el dorado, *Salminus maxillosus*. Determinar los niveles de cosecha sostenible y ajustar la legislación de acuerdo de lo que se encuentra. Regular la cosecha de los peces. La cooperación internacional debe realizarse porque los peces cruzan las fronteras entre países.

- Tal como está propuesto, el Proyecto Hidrovía es una amenaza muy fuerte contra la biodiversidad de la cuenca del Río Paraguay. La reducción del número de peces en la región tendrá consecuencias muy importantes en el campo socio-económico (e.g,. se perdería o reduciría la pesca comercial) además de bajar el valor esencial de la biodiversidad. Si el Proyecto Hidrovía se lleva a cabo tal como está previsto, se debe planificar de antemano un refugio adecuado dentro del cual la biodiversidad puede sostenerse. Las playas del Río Paraguay es un macrohábitat que funciona como fuente para otros macrohábitats en donde la población de peces podrían reproducirse. Se puede hacer criaderos de algunos organismos, como los cangrejos y los camarones, con una facilidad relativa. La mayoría de las especies pueden ser colectadas y mantenidas como "banco de recolonización" para cuando se las necesite. Estas actividades ayudarían a la preservación de las poblaciones de Río Paraguay, las que son únicas en el mundo.

- Continuar las investigaciones de la flora y la fauna de la región. Tanto los estudios ecológicos como los sistemáticos serían muy valiosos. Estudios hidrológicos también son necesarios.

- Peces para acuarios (e.g., cichlidium, characidium, loricariids), fueron colectados durante la expedición, en las playas del Río Apa y microhábitats de agua clara en particular. La cosecha a gran escala y no restringida destruirían estas poblaciones, como ya ha ocurrido en otras áreas de Sudamérica. Luego de determinar los tamaños de poblaciones y de hacer los estudios de impacto poblacional, puede ser posible que los pescadores locales implementen la pesquería sostenible de estas especies de peces utilizadas en los acuarios.

Literatura Citada

Bonetto, A.A. 1986a. The Paraná River system. *In* Davies, B.R. and K.F. Walker (eds.). The ecology of river systems. Dordrecht, Netherlands: Dr. W. Junk Publishers. Pp. 541–555.

Bonetto, A.A. 1986b. Fish of the Paraná system. *In* Davies, B.R. and K.F. Walker (eds.). The ecology of river systems. Dordrecht, Netherlands: Dr. W. Junk Publishers. Pp. 573–588.

Chernoff, B., P.W. Willink, M. Toledo-Piza, J. Sarmiento, M. Medina, and D. Mandelburger. 2001. Testing hypotheses of geographic and habitat partitioning of fishes in the

Río Paraguay, Paraguay. *In* Chernoff, B., P. W. Willink, and J.R. Montambault, (eds.). A biological assessment of the aquatic ecosystems of the Río Paraguay basin, Departamento Alto Paraguay, Paraguay. RAP Bulletin of Biological Assessment No. 19. Washington, DC: Conservation International. Pp. 80–98.

Ferreira, C.J.A., B.M.A. Soriano, S. Galdino, and S.K. Hamilton. 1994. Anthropogenic factors affecting waters of the Pantanal wetland and associated rivers in the upper Paraguay basin of Brazil. *In* Barbosa, F.S.R. (ed.). Acta Limnologica Brasiliensia, volume V, workshop: Brazilian programme on conservation and management on inland waters. Fundacão Biodiversitas—Sociedade Brasileira de Limnologia, Belo Horizonte, Brasil. Pp. 135–148.

Gleick, P. H. 1998. The world's water. The Biennial Report on Freshwater Resources. Washington, DC: Island Press. 308pp.

Goulding, M. 1980. The fishes and the forest. Berkeley: University of California Press. 280pp.

Heckman, C.W. 1998. The Pantanal of Poconé: biota and ecology in the northern section of the world's largest pristine wetland. Boston: Kluwer Academic Publishers. 622pp.

Lowe-McConnell, R. H. 1975. Fish communities in tropical freshwaters: their distribution, ecology and evolution. New York: Longman. 337pp.

Lourival, R. F. F., Silva, C. J, Calheiros, D. F., Albuquerque, L. B., Bezerra, M. A. O., Boock, A., Borges, L.M.R., Boulhosa, R.L.P., Campos, Z., Catella, A. C., Damasceno, G. A. J., Hardoim, E.L., Hamilton, S.K., Machado, F.A., Mourão, G. M., Nascimento, F.L., Nougueira, F.M.B., Oliveira, M.D., Pott, A., Silva, M.C., Silva, V.P., Strussmann, C., Takeda, A. M. and Tomás, W. M. (1999). Os impactos da Hidrovia Paraguai-Paraná sobre a biodiversidade do Pantanal - Uma discussão multidisciplinar. *In* Anais do II Simpósio sobre recursos naturais e sócio-económicos do Pantanal. Corumbá - MS, Brasil, 517–534 pp.

Lourival, R.F.F. et al. 1996. Os impactos da Hidrovia Paraguai-Paraná sobre a biodiversidade do Pantanal— uma discussão multidisciplinar. *In* Anais do II Simpósio sobre Recurusos Naturais e Sócio-econômicos do Pantanal. Corumbá, MS, Brasil. 497pp.

Lundberg, J.G., L.G. Marshall, J. Guerrero, B. Horton, M.C.S.L. Malabarba, and F. Wesselingh. 1998. The stage for neotropical fish diversification: a history of tropical South American rivers. *In* Malabarba, L.R., R.R. Reis, R.P. Vari, Z.M.S. Lucena, and C.A.S. Lucena (eds.). Phylogeny and classification of neotropical fishes. Porto Alegre, Brasil: EDIPUCRS. Pp.13–48.

Moore, D. 1997. Hidrovia studies called profoundly flawed. World Rivers Review 12. 3pp.

Nakatani, K., G. Baumgartner, and M. Cavichioli. 1997. Ecologia de ovos e larvas de peixes. *In* Vazzoler, A. E. A. M., A. A. Agostinho, and N. S. Hahn (eds.). A planície de inundação do alto Rio Paraná: aspectos físicos, biológicos e socioeconómicos. Brasil: Editora da Universidade Estadual de Maringa. Pp. 281–306.

Pearson, N.E. 1937. The fishes of the Beni-Mamoré and Paraguay basins, and a discussion of the origin of the Paraguayan fauna. Proceedings of the California Academy of Science, 23:99–114.

Roberts, T.R. 1972. Ecology of fishes in the Amazon and Congo basins. Bulletin of the Museum of Comparative Zoology, 143:117–147.

Sowls, Lyle. 1997. Javelinas and other peccaries: their biology, management, and use. 2nd Ed. Arizona: Texas A & M University Press.

Switkes, G. 1996a. Belated public access to Hidrovia studies allowed. World Rivers Review, 10. 1pp.

Switkes, G. 1996b. Design chosen for first phase of Hidrovia. World Rivers Review, 11. 1pp.

Switkes, G. 1996c. Hidrovia plans propose heavy engineering for world's largest wetlands. World Rivers Review, 11. 2pp.

Switkes, G. 1997. Hydrologists condemn Hidrovia environmental studies. World Rivers Review, 12. 1pp.

Weil, T.E., J.K. Black, H.I. Blutstein, D.S. McMorris, F.P. Munson, and C. Townsend. 1984. Paraguay: a country study. Superintendent of Documents, Washington DC: U.S. Government Printing Office. 318pp.

WWF 2000. Retrato da Navegação no Alto Rio Paraguai. Relatório da Épedição Técnica Realizada entre os dias 3 e 14 de novembro de 1999 no Rio Paraguai entre Cáceres (MT) e Porto Murtinho (MS). Washington DC: World Wildlife Fund.

Sumário Executivo (Português)

Introdução

O Rio Paraguai é um rio de grande porte com muitos meandros e quando encontra o Rio Paraná, forma a segunda maior bacia fluvial da América do Sul (Mapa 1). O tamanho dessa bacia contradiz o fato do Rio Paraguai possuir águas relativamente rasas, descarregando somente 314 km³ de água por ano na bacia do Paraguai-Paraná. Este volume é pequeno se comparado ao da bacia do Río Orinoco, que deságua mais de quatro vezes esse volume (Gleick, 1998). O Rio Paraguai, que nasce nos Planaltos Mato-grossenses, flui através das áreas inundáveis do Pantanal e ao longo da borda leste do Chaco, dividindo o país do Paraguai (Mapa 1). O Rio Paraguai é um recurso aquático fundamental para as populações do Paraguai e do Brasil, não somente pela sua grande diversidade, mas também pela dependência humana dos recursos pesqueiros e da água doce para o consumo e agricultura.

A biodiversidade nas áreas de influência do Rio Paraguai representa uma transição interessante entre rios tropicais da Amazônia ao norte e hábitats temperados do sul. Atividades geológicas nos últimos 90 milhões de anos forneceram oportunidades para que os rios e seus habitantes mudassem da bacia Amazônica para a do Paraguai e vice-versa (Heckman, 1998:31–32; Lundberg et al., 1998). Acredita-se que a similaridade da flora e fauna aquáticas em relacó àquelas da Amazônia é consequência de intercâmbios nas nascentes entre os rios Guaporé e Arinos (ambos na Amazônia) e as nascentes do Rio Paraguai durante épocas de enchentes (Pearson, 1937; Roberts, 1972; Lowe-McConnell, 1975: 47, 85; Bonetto, 1986a, b). Conexões aquáticas tanto no passado quanto no presente ajudam a explicar a presença de espécies no Paraguai que são freqüentemente encontradas ao longo da costa Atlântica, ou ao sul na Argentina e Uruguai. Quando somados a esses fatos o endemismo, conclui-se que a biota do Rio Paraguai é única (Heckman, 1998: 31–33).

A caracterização da flora e fauna é um passo fundamental na documentação da diversidade aquática e da situação ambiental do ecossistema aquático, de modo que estratégias de conservação e desenvolvimento sustentável possam ser desenvolvidas na região. As ameaças aparecem em muitas formas, mas a principal causa de preocupação é o Projeto Hidrovia (Paraguai-Paraná). A hidrovia é um projeto que pretende transformar 3.400 km do rios Paraguai e Paraná navegáveis para barcas e grandes navios através de dragagem, correção de curvas, escavação de novos canais e destruição de afloramentos rochosos. A área afetada iria de Cáceres, Mato Grosso, no Rio Paraguai, rio-abaixo atingindo a foz do Rio Paraná no Oceano Atlântico (Mapa 1). Muitas das modificações fluviais ocorreriam no Pantanal (Lourival et al., 1999; Switkes, 1996a, b, c; WWF, 2001) com a intenção de facilitar o transporte, principalmente de soja e minério de ferro, durante todo o ano. Argentina, Uruguai, Paraguai, Bolívia e Brasil estão envolvidos nesse projeto (Moore, 1997; Heckman, 1998: 497).

As conseqüências ambientais do projeto incluiriam o escoamento e transformação de grandes extensões do Pantanal em áreas secas, diminuição do volume de enchentes nas áreas próximas da nascente do Rio Paraguai e aumento do volume de enchentes nas áreas próximas da foz (Switkes, 1996b, c, 1997). A salinização de alguns tributários, contaminação da água potável e o aumento no risco de derramamentos de óleo devido ao aumento do tráfego, são conseqüências prováveis desse projeto. Acredita-se que os benefícios econômicos tenham sido superestimados, pois o tráfego rodoviário existente e a competição potencial com as estradas de ferro não foram levados em consideração (Switkes, 1996c, 1997; Moore, 1997; Heckman, 1998: 497–498).

A destruição e modificação do hábitat não estão restritas ao Projeto Hidrovia. A população do Paraguai tem aumentado, em conjunto com a agricultura, silvicultura e pecuária. Essas atividades são geralmente mais intensas nas proximidades de áreas de colonização e vias de transporte, como estradas e hidrovias. A agricultura e pecuária envolvem queimadas e desmatamento de vastas porções de terra. Quando irrestritas e conduzidas por um período prolongado, essas modificações ambientais ameaçam a biodiversidade em determinadas áreas e aumentam a erosão.

A Expedição AquaRAP

Em resposta ao Projeto Hidrovia e outras ameaças, uma equipe multidisciplinar e multinacional de cientistas embarcou em uma expedição de Levantamento Rápido de Ecossistemas Aquáticos (AquaRAP) ao longo do Rio Paraguai com os seguintes objetivos: (i) descrever as características biológicas e ambientais do ecossistema aquático antes da implementação de qualquer modificação em grande escala e (ii) determinar, se possível, as modificações potenciais nos recursos aquáticos em resposta às ameaças recebidas. A expedição foi realizada entre os dias 4 e 18 de setembro de 1997 e amostrou a porção do Rio Paraguai entre os rios Negro ao norte e Aquidabán ao sul (Mapa 1). Partes do Rio Apa e do tributário Riacho La Paz também foram pesquisadas. A área estudada foi dividida em cinco regiões de forma a permitir comparações: Rio Negro, Alto Rio Paraguai (à montante do Morro Pão de Açúcar, aproximadamente 21° 26'S, 57° 55' W), baixo Rio Paraguai (à jusante do Morro Pão de Açúcar), Rio Apa e Riacho La Paz (Mapa 1). Foram estudados a vegetação terrestre e aquática, as características físicas e químicas da água, o plâncton, os macroinvertebrados aquáticos e os peixes. Esses inventários serão utilizados como base para a conservação e futuras recomendações de pesquisa na área, levando-se em consideração o Projeto Hidrovia e outros fatores.

O Programa de Levantamento Rápido de Ecossistemas Aquáticos da América do Sul é multinacional, multidisciplinar e voltado à identificar prioridades de conservação e oportunidades de manejo sustentável para os ecossistemas de água doce na América Latina. A missão do AquaRAP é identificar o valor biológico e de conservação dos ecossistemas de água doce através de inventários rápidos, e de divulgar as informações rapidamente para órgãos locais de fiscalização, conservacionistas, cientistas e agências internacionais de fomento. O AquaRAP é um programa gerenciado pela Conservation International em colaboração com o Chicago Field Museum.

O núcleo do AquaRAP consiste de um comitê diretor internacional composto por cientistas e conservacionistas de sete países (Bolívia, Brasil, Ecuador, Paraguai, Peru, Venezuela e Estados Unidos). O comitê diretor supervisiona os protocolos de levantamento rápido e a indicação de locais prioritários para tais levantamentos. As expedições do AquaRAP, que envolvem extensas colaborações com cientistas dos países onde elas se realizam, também promovem a troca de informações e oportunidades de treinamento em um âmbito internacional. As informações obtidas nas expedições do AquaRAP são divulgadas através do *RAP Bulletin of Bio-*

logical Assessment da Conservation International, projetado para administradores, políticos, líderes e conservacionistas que podem definir prioridades de conservação e direcionar ações através de financiamentos na região.

Uma avaliação da biodiversidade atual nos rios Paraguai e Apa se torna crucial antes que modificações em larga escala aconteçam na região. Essa parte do Paraguai tem sido lamentavelmente sub-amostrada, sendo que a maioria das coleções feitas no passado vem da vizinhança de grandes centros populacionais. Informações adicionais são necessárias para permitir a identificação de oportunidades de manejo. A seção seguinte resume as observações feitas durante a expedição AquaRAP. Para informações mais detalhadas, consulte os textos integrais dos capítulos. Após a síntese mencionada acima, apresenta-se um resumos dos capítulos, seguido das recomendações de atividades de conservação e pesquisa.

Sumário Geral

A expedição AquaRAP proporcionou um melhor conhecimento da região: antes dessa expedição ele se restringia a inventários florestais, estudos hidrológicos e de salinização (Rio Verde), alguns estudos da vegetação litorânea e coletas esporádicas de peixes. O AquaRAP compilou simultaneamente dados botânicos, limnológicos, de invertebrados e ictiológicos. A integração desses dados forneceu uma visão abrangente do ambiente aquático. É com base nessa visão que fazemos as observações e recomendações abaixo.

Rio Paraguai

O extremo norte do Rio Paraguai, no país do Paraguai, se localiza entre dois hábitats que não são encontrados em nenhuma outra parte do mundo, o Pantanal e o Chaco. O Pantanal é a maior área inundável do mundo, renomado pelos seus incríveis bandos de aves e grande número de jacarés, capivaras, veados, tamanduás e outros animais de grande porte. A paisagem aberta faz dele o paraíso dos observadores de animais. O Chaco é uma vasta planície de beleza singular, dominada por gramíneas e arbustos cerrados. A água é escassa, sendo fornecida somente pelas chuvas e enchentes sazonais. A fauna do Chaco inclui onças, emas, veados, lobos-guarás, sucuris e outras serpentes, como as cascavéis, além de tamanduás, antas, porcos-do-mato (incluindo o taguá, que acreditava-se estar extinto) e grupos de aves, muitas delas migratórias e similares àquelas do Pantanal. Esses animais tornaram-se adaptadas às condições árduas e são surpreendentemente numerosos. A abundância e diversidade de organismos é mais alta ao longo de rios e pântanos (Weil et al., 1984: 17–19). A singularidade global do Pantanal e do Chaco e a sua dependência de fenômenos hidrológicos específicos, torna-os adequados à iniciativas de conservação do ecossistema aquático, dado que essas regiões não foram severamente destruídas por atividades antrópicas.

A área estudada pela expedição AquaRAP está atualmente em boas condições ambientais. Os moradores locais dependem muito da água para sobrevivência, mas os níveis populacionais humanos e o número de fazendas são baixos. A produtividade é alta, com macrófitas aquáticas sendo responsáveis pela maioria da produção primária e decomposição que traz nutrientes para o rio. A pesca é sub-utilizada e nenhuma contaminação direta da água foi observada. Infelizmente, no entanto, essa situação provavelmente mudará.

Há expectativa de crescimento populacional. Existe espaço atualmente para o desenvolvimento na região, mas o crescimento desordenado poderá causar problemas a longo prazo. Estudos de impacto são fundamentais na determinação de como manter fazendas e a pesca em níveis sustentáveis. Novas fazendas agrícolas e pecuárias aumentarão a erosão e resultarão na perda de espécies valiosas, como já está acontecendo próximo à Bahia Negra. É provável que ocorra poluição comercial e residencial da água, com subseqüente eutrofização e problemas de saúde. Precauções, tais como a manutenção da vegetação ciliar para limitar a erosão e o tratamento da água utilizada, precisam ser tomadas de modo a evitar essas conseqüências negativas. A pressão da pesca comercial e turística poderá aumentar e as taxas de reprodução e crescimento dos peixes envolvidos devem ser consideradas quando a legislação tiver sendo elaborada. Se a dinâmica populacional for ignorada, será somente uma questão de tempo antes que os recursos pesqueiros entrem em colapso. No geral, pesquisas adicionais sobre a ecologia da região são necessárias para determinar os níveis sustentáveis de agricultura e pesca.

A maior ameaça ambiental provavelmente vem do projeto da Hidrovia Paraguai-Paraná, que objetiva fazer do Rio Paraguai uma via navegável para navios de porte oceânico. O impacto exato do Projeto Hidrovia é desconhecido (veja Mapa 1). Prevê-se que partes das áreas inundáveis do Pantanal próximas às cabeceiras do Rio Paraguai irão secar e que as enchentes em direção à foz irão aumentar (Lourival et al., 1999; Switkes, 1996b,c, 1997; WWF, 2001). É difícil predizer a dimensão do impacto hidrológico na área de estudo mas, quaisquer que sejam essas conseqüências, a distribuição e composição das comunidades aquáticas atuais seriam alteradas. O simples fato dos danos resultantes da dragagem e canalização não poderem ser, ou não estarem sendo estimados, constitui razão suficiente para preocupação.

A diminuição das enchentes e a resultante perda de áreas alagáveis seria devastadora para as comunidades aquáticas. O fenômeno ecológico dominante na região é a enchente sazonal, com organismos aquáticos e semi-aquáticos adaptados às inundações periódicas. Essas inundações redistribuem os nutrientes, facilitando o crescimento da vegetação nas áreas alagáveis. Muitas plantas também dependem das enchentes para dispersar suas sementes (Goulding, 1980). O aumento do nível da água serve como um sinal para a migração reprodutiva de várias espécies comercialmente valiosas de peixes e, para muitos organismos, a maioria do crescimento anual acontece durante períodos de cheia (Lowe-McConnell, 1975; Goulding, 1980; Bonetto, 1986b). Áreas inundadas servem de berçário e abrigo para muitas espécies de peixes e invertebrados (Bonetto, 1986b; Nakatani et al., 1997). A interrupção das enchentes diminuiria dramaticamente a heterogeneidade, biodiversidade e biomassa do hábitat.

Os efeitos da Hidrovia Paraguai-Paraná não seriam limitados aos ecossistemas aquáticos. A abundância de peixes diminuiria, com efeitos desastrosos tanto para a pesca, quanto para as aves, mamíferos e répteis que dependem de peixes na alimentação. O aumento das inundações à jusante destruiria plantações existentes, forçando o deslocamento das populações ribeirinhas.

Rio Apa

O Rio Apa encontra-se atualmente em boas condições ambientais. A qualidade da água é alta e os níveis populacionais humanos são relativamente baixos. A agricultura, pecuária e silvicultura são atividades presentes na região. O rio fornece água limpa e alimento para comunidades locais e para várias aves, répteis e mamíferos.

Apesar do baixo Rio Apa ser similar ao Rio Paraguai, a parte alta do rio é distinta ecologicamente, com margens íngremes e bancos de areia. Essas diferenças no hábitat fazem com que a fauna e flora sejam distintas. Devem ser tomadas medidas no sentido de reservar áreas nas porções alta e baixa do rio. A heterogeneidade do hábitat e a biodiversidade da região toda podem ser mantidas através da conservação de amostras representativas dos rios Apa e Paraguai.

Com seus bancos íngremes e solo enfraquecido, o Rio Apa se mostra particularmente susceptível à erosão. A vegetação ciliar protege grande parte do solo. Essa medida efetiva e natural anti-erosão é ameaçada por desmatamentos e a agricultura de coivara, pois ambos removem a vegetação.

A erosão pode causar problemas múltiplos. Um deles é a perda do valioso solo superficial, por fim resultando na diminuição da produtividade agrícola, que então teria de ser mantida através do uso de fertilizantes. Esses fertilizantes seriam escoados para o rio, resultando em eutrofização. Outro problema é a sedimentação do rio. A erosão em excesso faria com que o Rio Apa se tornasse túrbido, não mais possuindo hábitats singulares e, em conseqüência, reduzindo a heterogeneidade do hábitat e a biodiversidade regional.

A falta de regulamentação é o principal fator que faz com que a agricultura de coivara e a retirada de madeira sejam ameaças. Se medidas que limitem a destruição do hábitat, principalmente ao longo do rio, não forem tomadas, problemas de sedimentação e erosão irão ocorrer a longo prazo. A regulamentação dos tipos de árvore que podem ser retirados garantiria o sucesso econômico da região a longo prazo e preveniria a extinção de madeiras valiosas, como a *Amburana caerensis*. Com um plano de manejo adequado, talvez seja possível manter a produtividade e ao mesmo tempo preservar a integridade biológica do ecossistema aquático.

O Projeto Hidrovia impactaria o baixo Rio Apa de uma maneira similar àquela discutida acima para o Rio Paraguai. As conseqüências hidrológicas no alto Rio Apa não são claras pois esta região se encontra há uma altitude superior à do Rio Paraguai. Existe a possibilidade de que o lençol freático drene mais rapidamente, reduzindo o tamanho do rio sem aumento no lençol. Nesse caso, as populações de organismos aquáticos diminuiriam, mas não esperamos que os tipos de hábitat e a diversidade se alterem, a não ser que o volume de fluxo diminuísse substancialmente. O aumento das enchentes não está previsto para o alto Rio Apa.

Sumário dos Capítulos

Contexto Geográfico

O Rio Paraguai tem 2.550 km de comprimento e sua bacia cobre 1.100.000 km^2 (Bonetto, 1986a). Ele nasce na Chapada dos Parecis, Mato Grosso e em seguida flui em direção sul ao longo do Pantanal, a maior área inundável do mundo (Mapa 1). Depois de sair do Pantanal, o Rio Paraguai divide o país do Paraguai em duas metades, uma leste e outra oeste, antes de se juntar ao Rio Paraná. Juntos, os rios Paraná e Paraguai formam a segunda maior bacia hidrográfica da América do Sul.

Apesar dos níveis de água dos rios dependerem da quantidade de chuvas, a relação entre essas duas variáveis não é tão simples assim. No Pantanal, grande quantidade de chuvas ocorre durante os meses do verão, mas a pouca inclinação do terreno e o enchimento das áreas inundáveis mantêm a água no Brasil. Essa água é liberada aos poucos, indo em direção ao sul. De acordo com Ferreira et al. (1994), os picos do nível de água ocorrem na bacia do Rio Paraguai, em média, da seguinte maneira: no Pantanal do norte - de fevereiro a março; em Corumbá—de abril a maio; e na foz do Rio Apa—de julho a agosto. Esse fenômeno hidrológico atenua a magnitude das cheias à jusante. Apesar do nível da água poder alcançar até 4 metros no Pantanal, o Rio Paraguai sobe somente 1,5 metros em Assunção, no Paraguai (Weil et al., 1984:15).

O Rio Paraguai é um rio de grande porte com muitos meandros no norte do Paraguai, escoando através de um terreno relativamente plano. As flutuações nos níveis de água em conjunto com o terreno plano resultam em enchentes extensas e mudanças freqüentes no cursos dos rios. Braços mortos e canais abandonados são comuns e as margens do rio são ocupadas por florestas e carandazais. Esses ficam rotineiramente inundados durante muitos meses a cada ano. Grandes massas de vegetação flutuante (e.g., *Eichornia* sp.), conhecidos como " balseiros" são comuns. Os afloramentos rochosos próximos ao Morro Pão de Açúcar formam um gargalo natural, diminuindo o fluxo d'água a jusante.

O Rio Apa difere do Rio Paraguai por ser um rio menor com cascatas e quedas d'água. O leito é predominantemente arenoso e as margens são altas e bem definidas. É comum o acúmulo de troncos e galhos de árvores. A diversidade do hábitat no Rio Apa é menor do que no Rio Paraguai e os níveis de água podem mudar rapidamente em resposta às chuvas.

O Riacho La Paz pode ser descrito como um curso d'água de cabeceira que alterna cascatas e poças. O leito do rio é arenoso e pedregoso e o seu nível de água pode mudar rapidamente em resposta às chuvas, assim como o do Rio Apa. A diversidade geral de hábitats é baixa.

Vegetação Terrestre

Doze transectos foram estabelecidos para o estudo da vegetação terrestre ao longo dos rios Paraguai e Apa (Mapa 2). Também foi realizada uma coleta geral das plantas não encontradas nos transectos. Foram identificadas 198 espécies de plantas no total. Existem cinco tipos básicos de hábitat ao longo do Rio Paraguai: carandazais de *Copernicia alba* inundados sazonalmente, carandazais de *C. alba* não-inundáveis, matas alteradas de *Schinopsis balansae* modificadas, florestas de morro e florestas de Chaco.

Carandazais inundados sazonalmente são muito comuns ao longo do alto Rio Paraguai. Eles são dominados por *C. alba*, mas arbustos, lianas e gramíneas também são abundantes. Os carandazais de *C. alba* não-inundáveis são geralmente queimados e transformados em pastagens. O resultado é um carandazal alterado com arbustos pioneiros abundantes que mudam esse tipo de formação, tais como: *Acacia caven*, *Prosopis ruscifolia* e *Mimosa pellita*, dentre outros. As florestas alteradas de *S. balansae* são compostas de uma camada do tipo arbustiva de *Coccoloba* aff. *guaranitica*, *Tabebuia nodosa*, *Ziziphus mistol*, *Pseudobombax* sp., *Copernicia alba*, *Caesalpinia paraguariensis* e *Prosopis ruscifolia*.

As florestas de morro estão fora do alcance das inundações e o solo drena rapidamente. A flora é portanto similar àquela encontrada nas áreas com pouca água. As árvores tendem a ser pequenas e as gramíneas efêmeras. As florestas de Chaco são semelhantes, com árvores que alcançam 9 a 12 metros de altura. O sub-bosque é dominado por arbustos espinhosos e gramíneas.

Existem três tipos básicos de hábitat ao longo do Rio Apa: vegetação ciliar, florestas semi-decíduas, e cerradão. A vegetação ciliar margeia o Rio Apa e é inundada durante as cheias. As florestas semi-decíduas podem ser divididas em três níveis: o mais alto é composto de árvores com 25 a 30 metros de altura; o intermediário possui árvores com altura entre 15 e 20 metros constituído de *Croton urucurana*, *Trichilia* sp. e *Astronium fraxinifolium;* e o sub-bosque constitui-se de poucos arbustos com menos de 1 metro de altura formando moitas. Muitas das espécies são decíduas, mas durante as chuvas, esse hábitat lembra uma floresta úmida. As formações de Cerradão estão mais distantes do rio, onde grandes árvores são raras.

Como um todo, a vegetação terrestre adjacente aos rios é muito influenciada pelas flutuações no nível da água e pela sazonalidade das chuvas. Queimadas e alterações do hábitat, principalmente para a pecuária, têm tido um impacto significativo na região. Uma evidência disso é o fato de que não foram encontradas áreas com vegetação primitiva durante a expedição. O desmatamento excessivo está acelerando a erosão, resultando em sedimentação massiva do Rio Paraguai.

Vegetação Aquática

A vegetação aquática foi amostrada em 31 localidades ao longo do Rio Paraguai e sete ao longo do Rio Apa (Mapa 3). Um total de 147 espécies foi identificado no Rio Paraguai e 50 no Rio Apa, perfazendo um total de 187 espécies de plantas aquáticas.

As margens inundadas do Rio Paraguai são dominadas por um "mosaico de vegetação". Os componentes desse "mosaico" são florestas de *Schinopsis balansae,* cerrado-carandazal de *Copernicia alba,* baixadas inundáveis e vegetação de margem. No Rio Paraguai foi enfocada a vegetação de margem. Essa vegetação foi sub-dividida em hábitats semilóticos, de margem, de brejo e de bancos de areia. As espécies foram categorizadas de acordo com o modo de vida: epífita, flutuante-livre, liana, brejo, semi-submersa enraizada, ou submersa.

Os hábitats com águas de fluxo lento são encontrados onde o rio inunda suas margens. Esses hábitats são muito mais diversos do que seria esperado aleatoriamente, e provavelmente são responsáveis pela maioria da diversidade da flora aquática na área de estudo. Essa alta diversidade pode ser relacionada à dinâmica dos níveis da água e dos ciclos de nutrientes. Os habitats de margem são relativamente pobres em termos de vegetação aquática, provavelmente devido ao movimento constante da água que impede o crescimento de muitas espécies.

Os brejos são altamente variáveis. Alguns deles, com pouco sedimento, podem ser melhor descritos como "baseiros". Outros, geralmente em áreas protegidas do vento e de correntes e em um estado mais adiantado de desenvolvimento, possuem uma quantidade significativa de substrato. O número de espécies em uma dada área é aproximadamente proporcional à quantidade de substrato.

Os hábitats de banco de areia são em geral menos ricos em espécies, provavelmente devido à baixa qualidade de nutrientes no solo. No entanto, as espécies de plantas encontradas são muito distintas e características desses locais.

A diversidade da flora aquática do Rio Apa é muito mais baixa do que a do Rio Paraguai. Isso é provavelmente consequência da alta velocidade da corrente, da falta de diversidade do hábitat, da preponderância da areia e do grande número de troncos e galhos no curso d'água. Os ambientes mais diversos são encontrados em alguns dos tributários.

Qualidade da Água e Macroinvertebrados Bentônicos

A qualidade da água foi examinada em 35 localidades (Fig. 4.1). As águas do Rio Paraguai são em geral levemente ácidas (pH 6,0–6,5), com baixos níveis de oxigênio (<6,0 mg/L), baixa condutividade elétrica (60–100 µS/cm) e temperaturas entre 24–27° C. As águas do Rio Apa são naturalmente um pouco alcalinas (pH 7,27–7,96), com níveis variáveis de oxigênio dissolvido (4,2-7,13 mg/L), alta condutividade elétrica (163–376 µS/cm) e temperaturas que variam entre 22–29° C.

As comunidades fitoplanctônicas são relativamente diversas, incluindo representantes de vários grupos de algas, tais como Chlorophyta, Euglenophyta, Chrysophyta, Bacillariophyta e Cyanophyta. Os macroinvertebrados bentônicos foram coletados em 33 localidades. Larvas de Chironomidae (Diptera) e Oligochaeta foram os grupos dominantes em 27 localidades, representando 52% e 35% dos organismos registrados. Odonata, Trichoptera, Ephemeroptera, Chaoboridae, Ceratopogonidae, Corixidae, Conchostraca, Planaria, Nematoda, Hirudinea, Copepoda, Ostracoda, Bivalvia e Gastropoda foram coletados com menor freqüência.

Dentre as larvas de Chironomidae, 26 taxa foram identificados. Alguns podem representar espécies não descritas. A diversidade de macroinvertebrados bentônicos é alta quando comparada à de outras bacias da América do Sul. A maioria dos gêneros são típicos de brejos herbáceos, poças, lagos e partes de córregos e rios com movimento lento. Áreas ricas em vegetação em decomposição exibiram uma baixa diversidade.

Como um todo, observamos poucas evidências de contaminação direta das águas. As únicas exceções foram em áreas próximas à Baia Negra, onde queimadas e desmatamentos poderão resultar em erosão e eutrofização.

Macroinvertebrados

A fauna de macroinvertebrados aquáticos foi pesquisada em 54 localidades de coleta ao longo da cabeceira do Rio Paraguai, entre a foz do Rio Negro e a foz do Rio Aquidabán, e ao longo da seção intermediária do Rio Apa. Os esforços de coleta foram direcionados para o estudo da fauna de crustáceos decápodas, mas outros grupos foram incidentalmente coletados. Os hábitats amostrados incluem praias fluviais, águas paradas, lagoas, corredeiras, planícies alagadas, vegetação flutuante, serrapilheira submersa e hábitats crípticos. As coletas foram somente qualitativas e realizadas com redes de arrastão e de imersão, armadilhas e manualmente.

Foram encontradas treze espécies de crustáceos decápodas: seis espécies de camarões Palaemonidae e uma de Sergestidae e seis espécies de caranguejo Trichodactylidae. A ocorrência de *Pseudopalaemon* sp. consiste no primeiro registro desse gênero na bacia do Rio Paraguai, sendo provavelmente uma espécie nova. O camarão *Macrobrachium amazonicum* e os caranguejos *Trichodactylus borellianus* e *Sylviocarcinus australis* foram as espécies mais comuns. A fauna de insetos aquáticos foi a mais rica em número de espécies, com 83 espécies em 34 famílias e dez ordens. Coleoptera, Hemiptera e Odonata foram coletados mais freqüentemente. Os moluscos são um grupo importante em termos de abundância e biomassa, com pelo menos sete espécies de Gastropoda e três de Bivalvia. Outros grupos menos representativos incluem os crustáceos perifíticos e parasitas (Conchostraca, Ostracoda, Amphipoda, Isopoda e Branchiura), anelídeos e platelmintos.

Peixes

Duas equipes de ictiólogos amostraram peixes em um total de 111 localidades (Mapa 3b). Durante a expedição do AquaRAP, 136 espécies de peixes foram coletadas no Rio Paraguai, 85 no Rio Apa e 35 no Riacho La Paz, perfazendo um total de 173 espécies.

Os Characiformes contribuem com 54% da fauna de peixes do Rio Paraguai. Eles pertencem à uma variedade de categorias tróficas, incluindo herbívoros, insetívoros, piscívoros, onívoros e os que se alimentam de matéria orgânica no fundo dos cursos d'água. Os Siluriformes (bagres) e Perciformes (principalmente ciclídeos) foram o segundo grupo mais abundante. As piranhas (*Serrasalmus* spp.) foram comuns e muitos peixes tinham marcas de mordidas de piranhas. Alguns peixes (e.g., pacus, anostomídeos) se alimentam de frutas e sementes e dependem das cheias. Florestas inundadas são muitas vezes utilizadas como berçários por muitas espécies de peixes, incluindo algumas importantes comercialmente. Muitos taxa (e.g., *Lepidosiren paradoxa, Synbranchus marmoratus,* Callichthyidae) possuem adaptações para respirar ar, que são importantes para a sobrevivência durante a estação seca. Somente os ovos sobrevivem a estação seca em várias espécies, conhecidas como peixes anuais (e.g., *Rivulus* sp.).

A diversidade de peixes foi mais baixa no Rio Apa do que no Rio Paraguai, provavelmente devido à baixa heterogeneidade do hábitat. Characiformes, Siluriformes e Perciformes foram ainda os taxa dominantes. Muitas espécies coletadas no Rio Apa não foram coletadas no Rio Paraguai, ressaltando a distinção dessas duas bacias. Muitas espécies de Siluriformes foram encontradas sob troncos flutuantes. Cardumes de *Salminus, Brycon e Prochilodus* foram observados migrando rio-acima.

Partição Geográfica e do Hábitat dos Peixes

As 173 espécies de peixes de água-doce coletadas durante a expedição do AquaRAP no Rio Paraguai foram analisadas para se determinar a existência de algum padrão de distribuição dentro da região coletada. A região foi dividida em 5 sub-regiões e em 10 macrohábitats. Duas hipóteses nulas foram testadas e rejeitadas: (i) que os peixes estão distribuídos aleatoriamente com relação à sub-região; e (ii) que os peixes estão distribuídos aleatoriamente com relação ao macrohábitat. Os resultados mostraram que um forte efeito sub-regional foi evidente, sendo que dois elementos de distribuição estavam presentes, consistindo de: (i) uma associação entre as sub-regiões do Rio Paraguai e do Rio Negro e (ii) uma associação entre as sub-regiões do Rio Apa e Riacho La Paz. A análise das distribuições com respeito aos macrohábitats também revelou dois componentes. O primeiro relaciona as faunas das praias e dos principais cursos d'água com os macrohábitats inundados durante cheias sazonais, como as florestas alagadas e lagoas. O grupo associado ao ciclo das inundações é responsável por mais de 75% dos peixes coletados. O segundo componente, composto de hábitats encontrados no Rio Apa e Riacho La Paz (e.g., águas claras, corredeiras, etc.), possui um limite relativamente nítido em relação ao Rio Paraguai, havendo uma substituição de mais de 50% da fauna. Esses resultados mostram que o cerne das áreas de conservação precisa incluir áreas onde o ciclo anual de enchentes não seja impedido e que a área de inundação não seja muito perturbada. Além disso, o Rio Apa e o Riacho La Paz representam áreas muito ameaçadas pelas altas taxas de conversão de terras e de aridez, e pelo fato da recolonização faunística a partir de fontes de água próximas não ser possível.

Congruência dos Padrões de Diversidade entre Peixes, Invertebrados e Plantas Aquáticas

Foram testadas hipóteses nulas com relação à distribuição aleatória das espécies em relação às sub-regiões e macrohábitats no Rio Paraguai, a partir de dados referentes a 131 espécies de macrocrustáceos e invertebrados bentônicos e 186 espécies de plantas aquáticas. Os padrões foram comparados aos resultados para distribuição de peixes apresentados por Chernoff et al. (2001). Os dados referentes aos invertebrados demonstram um padrão idêntico entre sub-regiões, como é evidente na distribuição dos peixes. Os resultados apóiam o reconhecimento de duas zonas: (i) a zona do Rio Paraguai contendo o alto e baixo Rio Paraguai e o Rio Negro; e (ii) a zona do Rio Apa, contendo o Rio Apa e o Riacho La Paz. Em todos os conjuntos de dados, a zona do Rio Paraguai possui riqueza maior de espécies do que a zona do Rio Apa. O limite entre as duas zonas é abrupto, fato esse apoiado também pelos dados botânicos. Somente 11 das 186 espécies de plantas foram encontradas em ambas as zonas. Não existe congruência nos padrões entre macrohábitats. Os dados de invertebrados e de plantas contêm muitos valores que não diferem da similaridade média encontrada aleatoriamente. Os dados de peixes apresentam um padrão inequívoco de hábitats dentro da zona do Rio Paraguai atrelado à inundação de hábitats durante o ciclo das cheias. O conjunto de dados referente aos invertebrados forma um grupo que é consistente com o princípio geral. Os dados botânicos demonstram uma relação entre hábitats de margem e areia que estão expostos à correntes maiores do que outros macrohábitats. As plantas encontradas em hábitats de água parada possuem poucas similaridades com outros macrohábitats. Com base nessas observações conclui-se que porções significativas de hábitat dentro de cada uma das duas zonas precisa ser preservada de modo a manter uma grande parte da biodiversidade.

Recomendações de Atividades de Pesquisa e Conservação

A justificativa para as recomendações das atividades de pesquisa e conservação se encontram no sumário executivo e nos capítulos desse relatório. Muitas dessas recomendações requerem cooperação internacional pois os cursos d'água se originam e fluem através de vários países, sendo que alguns organismos (e.g., peixes migratórios) cruzam fronteiras internacionais com freqüência. As recomendações estão listadas em ordem de prioridade.

- Conservar áreas tanto do Rio Paraguai quanto do Rio Apa e não somente um ou outro. As regiões do alto Rio Paraguai-Rio Negro e do Rio Apa-Riacho de La Paz devem ser enfatizadas, pois juntas elas representam uma parte significativa da diversidade regional. A área explorada pela expedição AquaRAP é especialmente atrativa para o estabelecimento de reservas, pois a população humana nesse local é menor do que aquela nas regiões do Rio Paraguai situadas mais ao sul, que são mais densamente povoadas. Deve ser estabelecida uma cooperação entre o Paraguai, Brasil e Bolívia, pois esses cursos d'água cruzam as fronteiras desses países.

- Prevenir distúrbios de grande e pequena escala no ciclo sazonal de cheias. A interferência ou redução drástica do ciclo de cheias e dos hábitats inundados resultará na perda de áreas alagáveis (i.e., hábitats para plantas e invertebrados aquáticos, áreas de berçário para peixes, etc.) e pode resultar numa perda de 50% das espécies de plantas aquáticas e peixes da bacia. O represamento, a dragagem, a correção de curvas e a eliminação de afloramentos rochosos poderá influenciar as inundações.

- Avaliar o impacto das fazendas agrícolas e pecuárias no meio ambiente. Determinar níveis sustentáveis de produção. Elaborar legislações de uso da terra e de agricultura de coivara, retirada de madeira e ocupação humana. Prevenir a retirada de madeiras ameaçadas pela exploração descontrolada.

- Conservar a vegetação ciliar. Essa medida ajuda a prevenir a erosão e manter a fonte de nutrientes para o meio aquático. Programas de reflorestamento devem ser implementados nas áreas ribeirinhas que já se encontram danificadas.

- Não existe congruência entre os peixes, invertebrados e plantas aquáticas em relação à distribuição entre macro-hábitats. Como resultado, amostras de todos os macro-hábitats devem ser preservadas para manter a maioria das espécies.

- Estimar as taxas de reprodução e crescimento de espécies comerciais de peixes. Determinar os níveis sustentáveis de exploração e adaptar a legislação de acordo com esses níveis. Uma cooperação internacional se faz necessária, pois os peixes cruzam as fronteiras.

- Da forma que foi proposto, o Projeto Hidrovia consiste numa grande ameaça à biodiversidade aquática da bacia do Rio Paraguai. A redução do número de peixes na região teria conseqüências sócio-econômicas importantes (e.g., perda ou redução da pesca) além do valor intrínseco da biodiversidade. Caso o Projeto Hidrovia seja implementado, devem ser planejados, com antecedência, refúgios onde a maioria da biodiversidade possa se sustentar. No caso dos peixes, as praias do Rio Paraguai servem como uma fonte a partir da qual a recolonização para outros macrohábitats se faz possível. Alguns organismos, como caranguejos e camarões, se reproduzem com certa facilidade em cativeiro. A maioria das espécies pode ser coletada e mantida como "estoques de recolonização" para o futuro. Isto ajudaria a preservar a singularidade das populações paraguaias.

- Continuar pesquisando a flora e a fauna da região. Estudos de sistemática e ecologia são especialmente valiosos, mas estudos hidrológicos também são necessários.

- Peixes de aquário (e.g., ciclídeos, caracídeos, loricariídeos) foram coletados durante a expedição, especialmente ao longo das praias do Rio Apa em macrohábitats de água clara. A captura irrestrita em larga escala dessas espécies destruiria suas populações, como aconteceu em outras áreas na América do Sul. Talvez a exploração sustentável desses peixes por pescadores locais seja possível, depois que o tamanho das populações seja determinado e estudos de impacto sejam executados.

Literatura Citada

Bonetto, A.A. 1986a. The Paraná River system. *In* Davies, B.R. and K.F. Walker (eds.). The ecology of river systems. Dordrecht, Netherlands: Dr. W. Junk Publishers. Pp. 541–555.

Bonetto, A.A. 1986b. Fish of the Paraná system. *In* Davies, B.R. and K.F. Walker (eds.). The ecology of river systems. Dordrecht, Netherlands: Dr. W. Junk Publishers. Pp. 573–588.

Chernoff, B., P.W. Willink, M. Toledo-Piza, J. Sarmiento, M. Medina, and D. Mandelburger. 2001. Testing hypotheses of geographic and habitat partitioning of fishes in the Río Paraguay, Paraguay. *In* Chernoff, B., P. W. Willink, and J.R. Montambault, (eds.). A biological assessment of the aquatic ecosystems of the Río Paraguay basin, Departamento Alto Paraguay, Paraguay. RAP Bulletin of Biological Assessment No. 19. Washington, DC: Conservation International. Pp. 80–98.

Ferreira, C.J.A., B.M.A. Soriano, S. Galdino, and S.K. Hamilton. 1994. Anthropogenic factors affecting waters of the Pantanal wetland and associated rivers in the upper Paraguay basin of Brazil. *In* F.S.R. Barbosa, (ed.). Acta Limnologica Brasiliensia, volume V, workshop: Brazilian programme on conservation and management on inland waters. Fundação Biodiversitas-Sociedade Brasileira de Limnologia, Belo Horizonte, Brasil. Pp. 135–148.

Gleick, P. H. 1998. The world's water. The Biennial Report on Freshwater Resources. Washington, DC: Island Press. 308pp.

Goulding, M. 1980. The fishes and the forest. Berkeley: University of California Press. 280pp.

Heckman, C.W. 1998. The Pantanal of Poconé: biota and ecology in the northern section of the world's largest pristine wetland. Boston: Kluwer Academic Publishers. 622pp.

Lowe-McConnell, R. H. 1975. Fish communities in tropical freshwaters: their distribution, ecology and evolution. New York: Longman. 337pp.

Lourival, R. F. F., Silva, C. J, Calheiros, D. F., Albuquerque, L. B., Bezerra, M. A. O., Boock, A., Borges, L.M.R., Boulhosa, R.L.P., Campos, Z., Catella, A. C., Damasceno, G. A. J., Hardoim, E.L., Hamilton, S.K., Machado, F.A., Mourão, G. M., Nascimento, F.L., Nougueira, F.M.B., Oliveira, M.D., Pott, A., Silva, M.C., Silva, V.P., Strussmann, C., Takeda, A. M. and Tomás, W. M. (1999). Os impactos da Hidrovia Paraguai-Paraná sobre a biodiversidade do Pantanal - Uma discussão multidisciplinar. *In* Anais do II Simpósio sobre recursos naturais e sócio-econômicos do Pantanal. Corumbá - MS, Brasil, 517–534 pp.

Lundberg, J.G., L.G. Marshall, J. Guerrero, B. Horton, M.C.S.L. Malabarba, and F. Wesselingh. 1998. The stage for neotropical fish diversification: a history of tropical

South American rivers. *In* L.R. Malabarba, R.R. Reis, R.P. Vari, Z.M.S. Lucena, and C.A.S. Lucena (eds.). Phylogeny and classification of neotropical fishes. Porto Alegre, Brasil: EDIPUCRS. Pp.13–48.

Moore, D. 1997. Hidrovia studies called profoundly flawed. World Rivers Review 12. 3pp.

Nakatani, K., G. Baumgartner, and M. Cavichioli. 1997. Ecologia de ovos e larvas de peixes. *In* A. E. A. Vazzoler, M., A. A. Agostinho, and N. S. Hahn (eds.). A planície de inundação do alto Rio Paraná: aspectos físicos, biológicos e socioeconómicos. Brasil: Editora da Universidade Estadual de Maringa. Pp. 281–306.

Pearson, N.E. 1937. The fishes of the Beni-Mamoré and Paraguay basins, and a discussion of the origin of the Paraguayan fauna. Proceedings of the California Academy of Science, 23:99–114.

Roberts, T.R. 1972. Ecology of fishes in the Amazon and Congo basins. Bulletin of the Museum of Comparative Zoology, 143:117–147.

Sowls, L. 1997. Javelinas and other peccaries: their biology, management, and use. 2nd Ed. Arizona: Texas A & M University Press.

Switkes, G. 1996a. Belated public access to Hidrovia studies allowed. World Rivers Review, 10. 1pp.

Switkes, G. 1996b. Design chosen for first phase of Hidrovia. World Rivers Review, 11. 1pp.

Switkes, G. 1996c. Hidrovia plans propose heavy engineering for world's largest wetlands. World Rivers Review, 11. 2pp.

Switkes, G. 1997. Hydrologists condemn Hidrovia environmental studies. World Rivers Review, 12. 1pp.

Weil, T.E., J.K. Black, H.I. Blutstein, D.S. McMorris, F.P. Munson, and C. Townsend. 1984. Paraguay: a country study. Superintendent of Documents, Washington DC: U.S. Government Printing Office. 318pp.

WWF 2000. Retrato da Navegação no Alto Rio Paraguai. Relatório da Epedição Técnica realizada entre os dias 3 e 14 de novembro de 1999 no Rio Paraguai entre Cáceres (MT) e Porto Murtinho (MS). Washington DC: World Wildlife Fund.

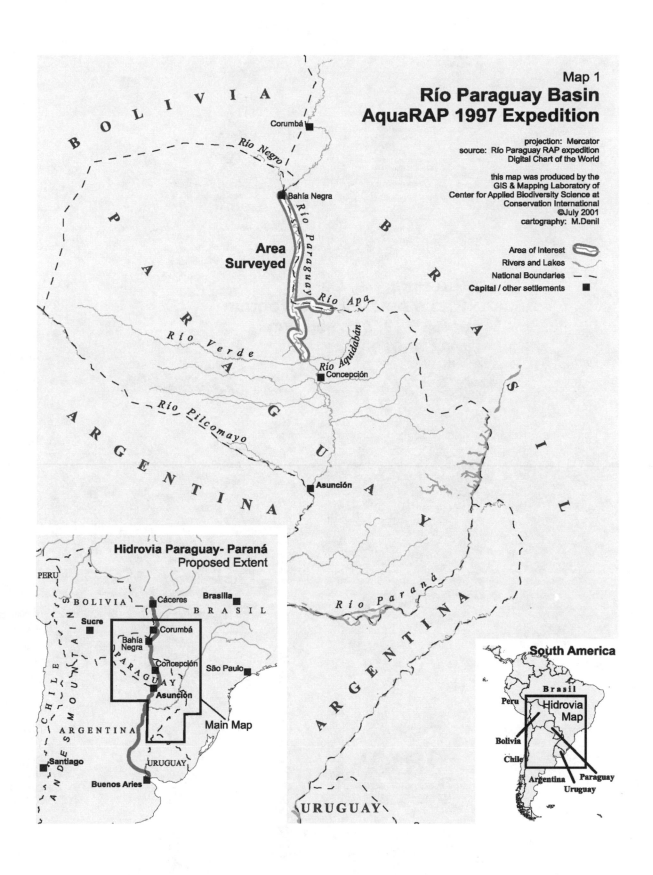

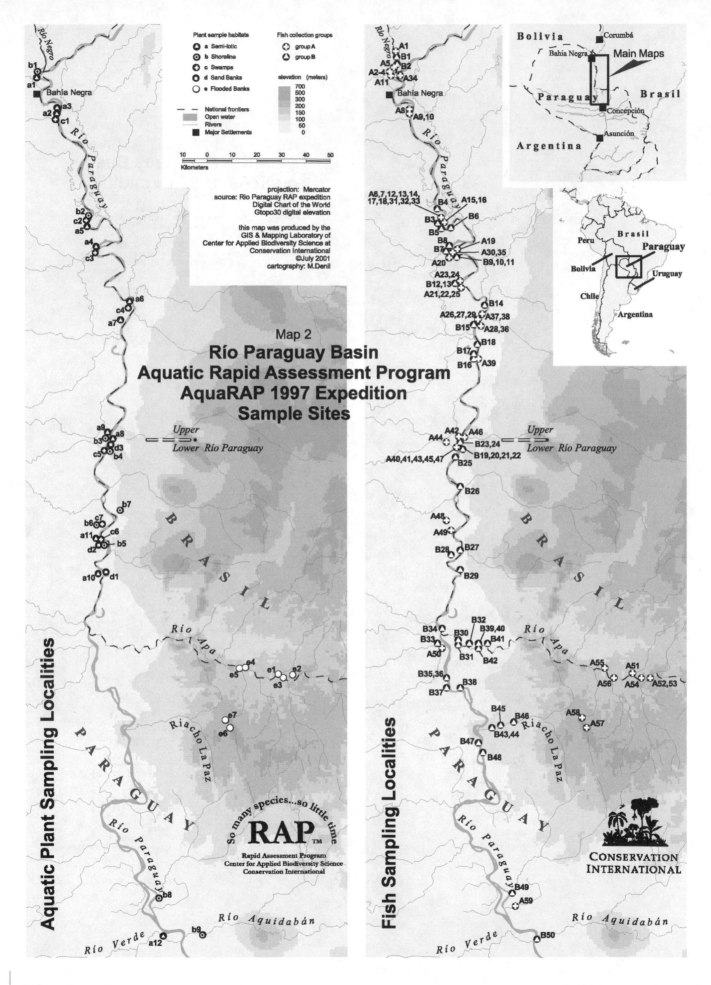

Map 2

Río Paraguay Basin
Aquatic Rapid Assessment Program
AquaRAP 1997 Expedition
Sample Sites

Aquatic Plant Sampling Localities

Fish Sampling Localities

Plant sample habitats
- a Semi-lotic
- b Shoreline
- c Swamps
- d Sand Banks
- e Flooded Banks

Fish collection groups
- group A
- group B

elevation (meters)
700
500
300
200
150
100
50
0

National frontiers
Open water
Rivers
Major Settlements

10   0   10   20   30   40   50
Kilometers

projection: Mercator
source: Rio Paraguay RAP expedition
Digital Chart of the World
Gtopo30 digital elevation

this map was produced by the
GIS & Mapping Laboratory of
Center for Applied Biodiversity Science at
Conservation International
©July 2001
cartography: M.Denil

Upper
Lower Río Paraguay

So many species...so little time
RAP™
Rapid Assessment Program
Center for Applied Biodiversity Science
Conservation International

CONSERVATION
INTERNATIONAL

Images of the AquaRAP Expedition

K. Awbrey

Liposarcus anisitsi: The skin of this armored catfish is covered with protective, armored plates. Although this individual is all black, this species, common in the aquarium trade, is usually colored with black-and-white reticulations, giving rise to the common name, Snow King. This fish can swim to the surface for air and can therefore withstand low oxygen levels. Like many catfish, *L. anisitsi* makes burrows or holes under water in muddy banks to live in and lay its eggs.

K. Awbrey

Serrasalmus spilopleura (above) and *S. marginatus* (below): Related to piranhas, this solitary species has very sharp teeth, which it uses to sneak up behind other fishes, including piranhas, and bite their fins or grab a mouthful of scales. Sometimes large piranhas will allow a *S. marginatus* to approach and pick off parasites. This cleaning behavior is more common, however, over coral reefs, and has rarely been documented in freshwater. Here, an adult and juvenile are pictured.

K. Awbrey

An AquaRAP botanist presses flowers in newspaper, a common field practice for preserving specimens.

K. Awbrey

Triportheus sp.: Commonly known as *pirá-guirá*, these fish are characids, known as tetras. With their long pectoral fins and keel-shaped chest, these fish have the agility to jump out of the water and evade predators. The flaps protruding from the fish's lower lip are extensions of the skin that help the fish utilize the oxygen-rich layer at the surface of the water. These flaps are only present when there is little oxygen in the water (e.g. at the end of the dry season/beginning of the rainy season).

Riparian vegetation and *Maurita* sp. palms along the Río Paraguay.

K. Awbrey

Dr. Jaime Sarmiento of the Museo de Historia Natural, La Paz, Bolivia, sets up fishnets for the ichthyology team on the Río Paraguay.

Satellite Image of the Río Paraguay Basin, Departamento Alto Paraguay, Paraguay

CHAPTER 1

Geographic Overview of the Río Paraguay Basin, Departamento Alto Paraguay, Paraguay

Philip W. Willink, Jaime Sarmiento,
Barry Chernoff, Mônica Toledo-Piza,
Darío Mandelburger, and Mirta Medina

Abstract

The Río Paraguay is 2,550 kilometers (1,580 miles) long and its basin covers 1,100,000 kilometers² (420,000 miles²; Bonetto, 1986a). It begins in the highlands of Planalto do Mato Grosso in southwestern Brasil, and subsequently flows south along the Pantanal, the world's largest wetland (Map 1). After exiting the Pantanal, the Río Paraguay splits the country of Paraguay into western and eastern halves before merging with the Río Paraná. Together, the ríos Paraguay and Paraná form the second largest watershed in South America.

Though water levels in the rivers depend upon the amount of rainfall, the relationship is not quite so simple. In the Pantanal, much of the rain falls during the summer months, but the low gradient and flooding of the wetlands maintains the water in Brasil. This water is slowly released and it moves as a pulse southward. According to Ferreira et al. (1994), peak water levels in the Río Paraguay basin occur, on average, in February to March in the northern Pantanal, April to May at Corumbá, and July to August at the mouth of the Río Apa. This hydrological phenomenon tempers the magnitude of downstream flooding. Even though water levels can change up to 4 meters (13 feet) in the Brazilian Pantanal, the Río Paraguay rises only 1.5 meters (5 feet) at Asunción (Weil et al., 1984:15).

The Río Paraguay in northern Paraguay is a large, meandering river which flows through relatively flat terrain. Water level fluctuations and the flat terrain result in widespread flooding and frequent changes in the course of the river. Oxbow lakes and abandoned channels are common. Forests and palm stands line the river's banks. These are routinely flooded for several months each year. Large masses of floating vegetation (e.g., *Eichhornia* sp.) known as floating meadows are common. Rock outcroppings near Cerritos Pão de Açúcar form a natural bottleneck, slowing the free flow of water downstream.

The Río Apa differs from the Río Paraguay in that it is a smaller river with cascades and waterfalls. The bottom is predominantly sand and the banks are high and well defined. Accumulations of tree trunks and branches are common. Habitat diversity in the Río Apa is lower than in the Río Paraguay. Water levels can change rapidly in response to rains.

The Riacho La Paz can be described as a headwater stream whose course alternates between riffles and pools. The bottom is sand and rock. Water levels can change rapidly in response to rains. Habitat diversity is low.

Introduction

To understand the biology of organisms, it is necessary to be familiar with the environmental variables to which they are, or potentially may be, exposed. This holds true for aquatic organisms, since geology, topography of the land, hydrology, climate, and other environmental features influence their distributions and life histories. This chapter will briefly describe some of the environmental features of the area surveyed by the 1997 AquaRAP expedition.

The Río Paraguay itself is 2,550 kilometers (1,580 miles) long, and its basin covers 1,100,000 kilometers² (420,000 miles²) (Bonetto, 1986a). It flows into the Río Paraná, which extends 4,000 kilometers (2,500 miles) and combines with the Río Paraguay to cover 2,800,000 kilometers² (1,100,000 miles²) with a yearly discharge of 500,000,000 meters³ (16,800,000,000 feet³) (Bonetto, 1986a). Together, the ríos Paraguay and Paraná form the second largest watershed in South America.

The Río Paraguay begins in the highlands of plan-alto of the southwestern state of Mato Grosso in Brasil. It almost immediately flows into the Pantanal, the world's largest wetland, which covers approximately 140,000 kilometers²

(54,000 miles^2), reaches 250 kilometers (155 miles) in width, and occupies portions of Brasil, Bolivia, and Paraguay (Bonetto, 1986b; Ferreira et al., 1994; Hamilton et al., 1997; Heckman, 1998:1). The Río Paraguay flows through elevations 80 to 150 meters (260 to 480 feet) (Ferreira et al., 1994), and the region is essentially the amalgamation of numerous alluvial fans. It is here in the Pantanal that the Río Paraguay grows in size due to the influx of the ríos Cuiab, São Lourenço, Taquarí, Negro, and Aquidauana among others. The majority of the Pantanal is encircled by the Serra dos Parecis, the Plan-alto of Mato Grosso, Serra de Maracajú, and other uplands with elevations exceeding 250 meters (800 feet) (Ferreira et al., 1994). A low point can be found south of Corumbá, Brasil, where the Río Paraguay enters the country of Paraguay.

The Río Paraguay forms the border between northern Paraguay and Brasil (Map 1). To the east can be found the southern vestiges of the Pantanal: specifically the Pantanal de Nabilique (Map 1). To the west is the Chaco, a flat, dry expanse with little in terms of topography (Weil et al., 1984:11). Streams, rivers, and swamps appear during the rainy season, but the majority disappear during the dry season. The swamps which remain are typically saline.

South of the Río Apa, the Río Paraguay divides the country of Paraguay into two regions. The east (known as Paraguay oriental) can be subdivided into two general regions. The eastern edge, which borders Brasil, is the Paraná Plateau. This plateau is flat, with many forested areas and serves as the area of origin for many of the eastern Paraguayan rivers, such as the Río Apa. These rivers flow westward, descend rapidly off the plateau, then slowly wind through the rolling countryside between the Paraná Plateau and the Río Paraguay. Over 90% of the human population in this area lives within Oriental Paraguay (Weil et al., 1984:11–12). The west (known as Paraguay occidental) is a damper portion of the Chaco where substantial swamps and several large, permanent rivers exist (Weil et al., 1984:11).

South of Asunción, where the Río Pilcomayo joins the Río Paraguay, the Río Paraguay forms the border between Argentina and southern Paraguay (Map 1). There is little topography in southern Paraguay, and swamps are common and widespread (Weil et al., 1984:12). At the southern tip of Paraguay, the Río Paraguay joins the Río Paraná. The Río Paraná continues southward through Argentina, eventually merging with the Río Uruguay to form the Río de La Plata which passes Buenos Aires, Argentina, and Montevideo, Uruguay, before emptying into the Atlantic Ocean.

Geology

Western Paraguay is a relatively flat expanse of land located between the Precambrian Brazilian Shield immediately to the east and the Late Cretaceous Andes distantly to the west.

From the Late Cretaceous to the Middle Miocene (approximately 90 to 12 million years ago), the exact courses of the rivers in Western Paraguay are difficult to reconstruct. The region was generally part of a proto-Paraná watershed or sea. At other times, the region likely drained into a paleo-Amazon-Orinoco basin, which emptied into what is now the Caribbean (Lundberg et al., 1998). During the Late Miocene (about 11 million years ago), an episode of Andean uplift resulted in subsidence of the land immediately to the east of the Andes. The Atlantic Ocean flowed into this depression, creating the Paranan Sea. This sea stretched to northern Argentina and possibly southern Bolivia, including western Paraguay. The ocean retreated around 9 million years ago (Marshall and Lundberg, 1996; Lundberg et al., 1998). Western Paraguay was subsequently covered by soil eroded from the Andes and the Brazilian Shield (Weil et al., 1984:20; Marshall and Lundberg, 1996; Horton and DeCelles, 1997). The region is now part of the Paraná watershed, with the Michicola Arch in Southern Bolivia forming the local basis for the drainage divide between the Amazon and Paraguay basins (Lundberg et al., 1998).

Climate and River Levels

Paraguay's climate can best be described as subtropical. Summer lasts from October to March, when the region's weather is dominated by warm, humid air coming from the Amazon basin. Cold, dry air from the south, called "surazos", dominate the winter weather (Weil et al., 1984:12–14; Killeen, 1998). In eastern Paraguay, temperatures can reach 40° C in the summer and dip below freezing in the winter, although monthly averages range from 18° to 29° C. Temperatures are always high in the Chaco, even during the winter (Weil et al., 1984:14).

Rainfall varies across the region, but is concentrated between October and March. The Chaco receives 500 to 1,000 millimeters (20 to 40 inches) a year; much of it evaporates immediately due to the high temperatures. Asunción has recorded 560 to 2,100 millimeters (22 to 82 inches), with the Paraná Plateau experiencing up to 3,300 millimeters (130 inches) in a single year (Weil et al., 1984:14–15). These are probably extreme values. The Pantanal receives about 1,100 millimeters (43 inches) annually (Schaller and Crawshaw, 1982; Heckman, 1998:45–49; but see Hamilton et al., 1997 for a slightly higher estimate). Typical yearly precipitation for the region surveyed by the AquaRAP expedition appears to be 1,100 to 1,400 millimeters (43 to 55 inches) (Heckman, 1998:48).

Though water levels in the rivers depend upon the amount of rainfall, the relationship is not quite so simple. In the Pantanal, much of the rain falls during the summer months, but the low gradient and flooding of the wetlands maintains the water in Brasil. This water is slowly released and it moves as a pulse southward. According to Ferreira et al. (1994), peak water levels in the Río Paraguay basin occur, on average, in February to March in the northern Pantanal, April to May at Corumbá, and July to August at the mouth of the Río Apa. This hydrological phenomenon tempers the magnitude of downstream flooding. Even though water levels can change up to 4 meters (13 feet) in the Brazilian Pantanal, the Río Paraguay rises only 1.5 meters (5 feet) at Asunción (Weil et al., 1984:15). This hydrological phenomenon helps temper the magnitude of downstream flooding. This provides graphic evidence for the interconnectedness and continuity of aquatic systems.

Although much of the land is flat, there are some places of higher elevation, such as near Cerritos Pão de Açúcar (21° 26' S, 57° 55' W). The rock outcroppings near this location act as a bottleneck; slowing the free flow of water downstream, causing the floodplain to be wider upstream from Cerritos Pão de Açúcar.

The combination of the pulse of water, generally low topography, and occasional rock outcroppings results in seasonal flooding that affects large areas adjacent to the Río Paraguay. The entire Pantanal, which stretches from near the headwaters of the Río Paraguay to the Río Apa, is flooded every year. Portions of the Chaco desert become swamps and flowing rivers (Weil et al., 1984:14).

Habitats of the Río Paraguay

Due to the size of the river and relatively large water level fluctuations in an area with flat terrain, the Río Paraguay creates a vast flooded plain that results in the formation of a high diversity of habitats.

Broad fringes of vegetation associated with the numerous tributaries (river branches, streams, backwaters) are found along the length of the main channel of the Río Paraguay. These tributaries can be, depending upon the season, of considerable widths and depths. Banks are poorly defined (only in some sections can one see the top part of the banks), and the bottom is basically unconsolidated clay-sand. Small sand beaches were found on rare occasions.

A characteristic element of the river is the presence of large areas occupied by tremendous quantities of floating vegetation like *Eichhornia* sp., floating Cyperaceas or Gramineas, and water hyacinth beds. The floating meadows occupy large areas and are generally found in slow currents, stream outlets, and small rivers, forming compact masses of floating vegetation that can obstruct navigation. The Río Paraguay is also characterized by vast seasonally flooded plains, creating flooded forests habitats and inundated palm stands. During the expedition, many of these areas were covered with more than 2 meters of water.

Due to intense fluvial dynamics, the river valley includes numerous abandoned branches and meanders that form

oxbow lakes of different sizes and maintain different degrees of communication with the principal river. Due to their fluvial origin, most of these lakes have a half moon shape. Frequently, because of increased settling of sediments, they have a lighter color than the main river and are often associated with, especially in shallower areas, floating or rooting herbaceous vegetation. These lakes are in different stages of succession, with some cases exhibiting abundant vegetation and others with vast surfaces of open water. The stage is dependent upon the degree of contact with the main river and their water supply. The depth depends on the level of succession, and can vary from a few meters to depths similar to the main channel.

Habitats of the Río Apa

The Río Apa basin includes numerous streams that originate in the lowlands and oxbow lakes with different degrees of connectivity to the main river. These lakes are in stages of varying succession related to their distance from the Río Apa. Generally within the study area, the Río Apa is a meandering river with well defined banks covered by a 20 meter high forest. The bottom is usually sandy and large sand beaches are common. The river is approximately 20–25 meters wide and 1 meter deep in the dry season. It is characterized by the alternation of deeper pools with slower moving water and shallower areas with more rapid currents. The river is usually deeper and the water moves slower in the meanders.

The Río Apa differs from the Río Paraguay by the presence of various waterfalls and rapids, such as the rapids upstream of San Carlos and a waterfall located near the outlet. The rapids constitute a 500 meter section with numerous rocks in a shallow area with a very strong current, making navigation difficult. The waterfalls downstream constitute an insurmountable obstacle for navigation that can only be circumvented through a side channel.

Furthermore, due to a shallower depth and a higher frequency of woods along the banks, there are commonly large accumulations of trees and branches in the river. These furnish an important habitat for many species of aquatic insects, crustaceans, molluscs, and fishes (primarily Siluriformes) that live in trunk cavities or among submerged branches.

In the Río Apa, rapid fluctuations in water levels, or flashfloods, can occur due to rains anytime of the year. Water drainage is faster than in the Río Paraguay because of the steeper inclination, which results in a reduced capacity for water accumulation. In the lower part of the Río Paraguay during the growth stage of a flashflood, a "water retention" phenomenon can occur, producing inverted current flows.

In general, the smaller system size, predominance of sand, and existence of a less developed flood plain causes a lower diversity of habitats in the Río Apa than in the Río Paraguay.

Habitats of the Riacho La Paz

Hydrologically, the Riacho La Paz can be described as a small basin which originates in a mountainous area and empties into the Río Paraguay. It flows for about 60 kilometers (38 miles) and its basin encompasses approximately 1,600 kilometers^2 (625 miles^2) (Map 1).

The upper parts of Riacho La Paz correspond to a minor fluvial system, with well-defined banks of consolidated sand and rocks. In the stream-bed, water flows over rocks alternating with small, round pebbles and sand. The channel is less than 10 meters wide, is generally less than 1 meter deep, and alternates between stretches of rapid currents in shallower areas and slower currents in deeper areas. Small logjams can be found that provide shelter for some species.

As with other systems, Riacho La Paz possesses a highly seasonal regime with a low water period from May to October–November, and a high water period from November to March. However, like other systems that originate in the nearby mountains, the regime is "flashier", with quick changes in water level in response to precipitation.

The upper part of the Riacho La Paz is characterized by low habitat diversity.

Literature Cited

Bonetto, A.A. 1986a. The Paraná River system. *In* Davies, B.R. and K.F. Walker (eds.). The ecology of river systems. Dordrecht, Netherlands: Dr. W. Junk Publishers. Pp. 541–555.

Bonetto, A.A. 1986b. Fish of the Paraná system. *In* Davies, B.R. and K.F. Walker (eds.). The ecology of river systems. Dordrecht, Netherlands: Dr. W. Junk Publishers. Pp. 573–588.

Ferreira, C.J.A., B.M.A. Soriano, S. Galdino, and S.K. Hamilton. 1994. Anthropogenic factors affecting waters of the Pantanal wetland and associated rivers in the upper Paraguay basin of Brazil. *In* Barbosa, F.S.R. (ed.). Acta Limnologica Brasiliensia, volume V, workshop: Brazilian programme on conservation and management on inland waters. Fundacão Biodiversitas—Sociedade Brasileira de Limnologia, Belo Horizonte. Pp. 135–148.

Hamilton, S.K., S.J. Sippel, D.F. Calheiros, and J.M. Melack. 1997. An anoxic event and other biogeochemical effects of the Pantanal wetland on the Paraguay River. Limnology and Oceanography, 42:257–272.

Heckman, C.W. 1998. The Pantanal of Poconé: biota and ecology in the northern section of the world's largest pristine wetland. Boston: Kluwer Academic Publishers. 622pp.

Horton, B.K. and P.G. DeCelles. 1997. The modern foreland basin system adjacent to the Central Andes. Geology, 25:895–898.

Killeen, T.J. 1998. Introduction. *In* Killeen, T.J. and T.S. Schulenberg (eds.). A biological assessment of Parque Nacional Noel Kempff Mercado, Bolivia. RAP Working Papers 10. Washington DC: Conservation International. Pp. 43–51.

Lundberg, J.G., L.G. Marshall, J. Guerrero, B. Horton, M.C.S.L. Malabarba, and F. Wesselingh. 1998. The stage for neotropical fish diversification: a history of tropical South American rivers. *In* Malabarba, L.R., R.R. Reis, R.P. Vari, Z.M.S. Lucena, and C.A.S. Lucena (eds.). Phylogeny and classification of neotropical fishes. Porto Alegre, Brasil: EDIPUCRS. Pp.13– 48.

Marshall, L.G. and J.G. Lundberg. 1996. Miocene deposits in the amazonian foreland basin. Science, 273:123–124.

Schaller, G.B. and P.G. Crawshaw, Jr. 1982. Fishing behavior of Paraguayan Caiman (*Caiman crocodilus*). Copeia, 1982:66–72.

Weil, T.E., J.K. Black, H.I. Blutstein, D.S. McMorris, F.P. Munson, and C. Townsend. 1984. Paraguay: a country study. Washington DC: Superintendent of Documents, U.S. Government Printing Office. 318pp.

CHAPTER 2

Terrestrial Flora of the Upper and Lower Río Paraguay Basin and the Río Apa Sub-basin, Paraguay

Hamilton Beltrán

Abstract

Twelve variable transects were established to assess the terrestrial vegetation along the Río Paraguay and the Río Apa (Map 2). General collections also were made of plants not encountered on the transects. A total of 198 plant species were identified. There were five basic types of habitats along the Río Paraguay: seasonally inundated palm stands of *Copernicia alba* (carandá'y), non-flooded palm stands of *C. alba*, modified forests of *Schinopsis balansae*, forested hills, and Chaco forests.

The seasonally inundated palm stands are very common along the Upper Río Paraguay. They are dominated by *C. alba*, but shrubs, lianas, and grasses are also abundant. The non-flooded palm stands of *C. alba* often are burned and converted into pasture. The result is a forest comprised almost solely of small *C. alba*, grasses, and cypresses. The modified forests of *S. balansae* were composed of a shrub-like layer of *Coccoloba* aff. *guaranitica*, *Tabebuia nodosa*, *Ziziphus mistol*, *Pseudobombax* sp., *Copernicia alba*, *Caesalpinia paraguariensis*, and *Prosopis ruscifolia*.

The forested hills are out of reach of the floods and the soil drains rapidly. The flora is therefore similar to that found in areas with little water. Trees tend to be small and grasses ephemeral. Chaco forests are similar, with trees reaching 9 to 12 meters in height. Spiny shrubs and grasses compose the undergrowth.

There are three basic types of habitats along the Río Apa: riparian vegetation, semideciduous forests, and cerradão. The riparian vegetation borders the Río Apa and is flooded during high waters. The semideciduous forests can be divided into three stories: the highest level is composed of trees 25 to 30 meters in height; the intermediate level is made up of *Croton urucurana*, *Trichilia* sp., and *Astronium fraxinifolium*; and the understory is merely a few bushes. Many of the species are deciduous, but during the rainy season this habitat would resemble a rain forest. Cerradão formations are furthest from the river, where larger trees are uncommon.

Overall, the terrestrial vegetation adjacent to the rivers is largely influenced by the fluctuations in water levels and the seasonality of the rains. Burning and habitat conversion, principally for cattle ranching, have had a significant impact on the region. Evidence for this is that pristine vegetation was not encountered during the expedition. Excess deforestation is accelerating erosion, resulting in major sedimentation of the Río Paraguay.

Methods

We used the variable transect method (R.B. Foster, N.C. Hernández E., E.K. Kakudidi, and R.J. Burnham, unpublished manuscript). A total of 12 transects were established (Appendix 1, Map 2). In each of them we counted 20 individuals. We established two categories, >5 centimeter diameter at breast height (DBH) and >15 centimeter DBH. We also recorded those with >30 centimeter DBH in some cases. General collections of plants that were not encountered on the transects were also made (Appendix 2). A total of 302 plant collections were deposited in the Herbario del Paraguay—Facultad de Ciencias Químicas (UNA) and The Field Museum in Chicago (FMNH).

Río Paraguay Floral Habitats

Seasonally Inundated Palm Stands of *Copernicia alba*

These are very common along the length of the Río Paraguay, from Bahía Negra to Los Cerritos. These palm stands occupy low plains areas on both sides of the river that are flooded for the majority of the year. It is sufficient for the river to rise only a few centimeters in order for the flooding to reach three or four kilometers inland. It appears that the small hills located in Los Cerritos act like natural containers of the river, with the bed of the river and subsequent flooding wider to the north and narrower to the south. Ecologically, this habitat is closely tied to other forms of aquatic life (zooplankton, fishes), which use this zone for feeding, refuge, and growth of juveniles.

This community is composed of a homogenous arboreal layer dominated by *Copernicia alba* (Transects 1, 2) (Appendix 1), with vigorous individuals reaching 10 to 12 meters in height. Occasionally, there are other trees like *Aporosella chacoensis*. The presence of shrubs, vines, and grasses increases the diversity and density. Additional important species encountered, besides the dominants, are various fabaceas (*Acacia* spp., *Caesalpinia* spp.), *Pacourina edulis*, *Mikania scandens*, *Phyla reptans*, and species of *Borreria*, *Cissus*, *Waltheria*, *Alternanthera*, *Combretum*, and *Echinodorus*.

Non-flooded Palm Stands of *Copernicia alba*
Some *C. alba* palm stands are not exposed to regular flooding. The vegetation is modified, and it may be better to call it savanna vegetation. These areas are burned and subsequently converted to pasture for cattle. The end result is a forest almost completely of *Copernicia alba*, with each individual small in size. In some cases, one encounters individuals of *Prosopis ruscifolia* (vinal) and *Prosopis vinalillo* (vinalillo). These are left to supply shade for the cattle. A profuse herbaceous vegetation of grasses and cypresses is also present.

Modified forests of *Schinopsis balansae*
These disturbed, shrub-like forests do not exceed 12 meters in height and are dominated by *Coccoloba* aff. *guaranitica*, *Tabebuia nodosa*, *Ziziphus mistol*, *Pseudobombax* sp., *Copernicia alba*, *Caesalpinia paraguariensis*, and *Prosopis ruscifolia*. This last species is common in disturbed environments. The undergrowth is dominated by various poaceas and species of caraguat, *Bromelia balansae* and *Aechmea distichantha*, among others.

The presence of "forest islands" surrounded by water should be noted. These are forests of *Schinopsis balansae*, which probably developed in depressed areas that are flooded when the river swells. Species that can endure periodic inundations are found here. Examples are *Tabebuia heptaphylla*, *Coccoloba* sp., *Cathormion polyanthum*, *Genipa americana*, and *Celtis pubescens*.

Forested hills
On the stretch of the Río Paraguay between Fuerte Olimpo and Los Cerritos, small promontories of rock (hills) of approximately 150 meters in height are easily distinguished from a distance. On these hills the rocks are exposed, the soil is shallow, and drainage is rapid. The hills are covered by vegetation adapted to long periods of little water (e.g.,

species of *Cnidosculos*, *Jatropha*, *Carica*). Here we encountered the first and only Cactaceae, *Stetsonia coryne*, of the entire expedition. The growth of grasses is ephemeral and dependent upon the seasonal rains. The height of the trees (5 centimeter DBH) does not surpass 12 meters (Transects 6, 7, 8) (Appendix 1). The dominant species are *Cordia glabrata*, *Astronium urundeuva*, *Stetsonia coryne*, and *Pseudobombax* sp. Infrequently, there are the spiny palms *Acrocomia totai* and a Myrtacea. There is a high species richness of deciduous shrub species, some of them spiny like *Celtis* sp., *Randia* sp., *Cnidosculos albomaculatus*, *Jatropha grossidentata*, and various Malvaceae.

Chaco Forests
This vegetation occupies flat areas that are not flooded. These forests are more frequent to the south, and are commonly deforested. Physiognomically, it is considered one unit. It is difficult to discern layers in a shrub forest with a height of 9 to 12 meters. One species does not dominate, but *Tabebuia nodosa* (labón) and *Pseudobombax* are more frequent. Between the lower stems of the trees grows a tangle of spiny shrubs with many shoots that make the forest difficult to walk through. Dominant species are *Coccoloba* "*glabra*" and *Coccoloba* "*pubescente*". The forest floor is covered by grasses and bromeliads (caraguat).

Río Apa Floral Habitats

Riparian vegetation
This vegetation is immediately above the high banks of the river and is flooded during the high-water season. Three vegetation levels are recognized. The first one is composed of *Inga marginata*, *Guadua* sp., *Ficus* sp., *Vitex megapotamica*, and *Astronium urundeuva*. The second level is composed of *Triplaris* aff. *guaranitica*, *Genipa americana*, *Croton urucurana*, *Cecropia* aff. *pachystachya*, *Acrocomia* sp., *Guazuma ulmifolia*, and *Crataeva tapia*. The third is composed of smaller species, such as *Salix humboldtiana*, *Rapanea* sp., *Attalea guaranitica*, and *Eugenia* sp.

The undergrowth is sparse with abundant lianas like *Combretum* sp., *Arrabidaea corallina*, and *Smilax* sp. Along tributaries, like the Arroyo Blandengue, the riparian forest is similar, with *Peltophorun dubium* in the first level, and *Hygrophylla* sp., *Cyperus* sp., *Heliotropium* sp., and *Eleocharis* sp. along the shore.

Semideciduous Forests
This type of vegetation is very close to the Río Apa. It is above sandy substrate that is topographically level with occasional depressions. The characteristic of this formation is that many species are deciduous, but in times of rain it would have certain structural similarity to the rain forests. One can clearly discern three levels: the highest with 25–30 meter

trees with luxuriously leafed branches and some emergents (*Vitex megapotamica, Hymenaea* sp., *Patagonula americana*), the intermediate of juveniles and medium trees (*Croton uru-curana, Trichilia* sp., *Astronium fraxinifolium*), and the lowest of a few bushes. In this type of forest, it is possible to ascertain that the majority of the diversity is due to tree species and that there has been little disturbance.

Cerradão

This vegetative formation is a little further from the river where the terrain is level and sandy. The dominant fauna is deciduous, with a canopy no more than 10 meters in height. Large trees are not common. Very dense formations are exclusively of *Eugenia* sp. and *Pouteria* sp. In the soil, there is no growth of herbaceous vegetation except for caraguatá (*Bromelia* spp.).

Recommended Conservation and Research Activities

- The basin of the Río Apa and nearby areas (principally Río Arriba de Puerto San Carlos) should be a high priority for conservation for the following reasons: the diversity of tree species is high, there have been few modifications to the ecosystem, and the soil is sandy and not fit for cattle ranching or agriculture.

- Continued removal of the vegetative cover will accelerate erosion, resulting in major sedimentation of the Río Paraguay. Steps to preserve riparian vegetation are needed. Reforestation programs are also encouraged.

- The terrestrial vegetation adjacent to the Río Paraguay and some effluents (e.g., the Río Aquidabán and the Río Apa) is largely influenced by seasonal fluctuations in water level. Habitat alterations that affect flooding will have a major impact on the native flora, and should be prevented.

- Many years of human intervention have modified to some degree the vegetative cover due to the inclination to burn the vegetation with subsequent introduction of cattle. Evidence for this is that pristine vegetation was not encountered during the expedition. Continued unregulated slash-and-burn agriculture is discouraged, and some areas should be set aside so they may return to their natural condition.

CHAPTER 3

Evaluation of the Aquatic Floral Diversity in the Upper and Lower Río Paraguay Basin, Paraguay

María Fátima Mereles

Abstract

Aquatic vegetation was sampled at 31 localities along the Río Paraguay and seven localities along the Río Apa (Map 2). A total of 147 species were identified along the Río Paraguay and 50 species along the Río Apa, for a grand total of 187 aquatic plant species.

The flooded banks of the Río Paraguay were dominated by a "mosaic of vegetation". The components of this "mosaic" are forests of *Schinopsis balansae*, savanna-palm stands of *Copernicia alba*, flooded lowlands, and shoreline vegetation. In the Río Paraguay, we focused on the shoreline vegetation. This vegetation was subdivided into semilotic, shoreline, swamp, and sand bank habitats. Species were categorized according to their life style: epiphytic, free floating, liana, marsh, rooted semi-submergent, or submergent.

Habitats with slow flowing water are found where the river overflowed its banks. These habitats are much more diverse than one would expect by chance, and probably account for the majority of aquatic floral diversity within the study area. This high diversity may be related to the dynamics of the water levels and nutrient cycles. Shoreline habitats are relatively poor in aquatic vegetation, possibly because the constant movement of water hinders the growth of many species.

Swamps are highly variable. Some, with little sediment, could better be described as "floating meadows". Others, usually in areas sheltered from the wind and currents and in more advanced stages of development, possess an appreciable amount of substrate. The number of species in a given area is roughly proportional to the amount of substrate.

Sand bank habitats are generally less rich in species, probably due to the poor quality of nutrients in their soils. However, the plant species encountered are very distinctive and characteristic of these sites.

The aquatic floral diversity of the Río Apa is much lower than that of the Río Paraguay. This is probably due to the high current velocity, lack of habitat diversity, preponderance of sand, and large numbers of trunks and branches throughout the channel. The more diverse environments are found in some of the tributaries.

Introduction

The Pantanal is an immense sedimentary plain that is altered constantly by periodic inundations. The northern portion in Brasil is called the Pantanal Mato-Grossense. The southern limit reaches to the vicinity of the mouth of the Río Negro and its confluence with the Río Nabileque. Others consider the Pantanal de Nabileque, which extends from the Río Negro to the Río Apa, to be part of the entire Pantanal complex (Cabrera and Willink, 1976; Adamoli, 1982). It is approximately 680 kilometers in a straight line from Cáceres, Brasil to the Río Apa.

The region of the Upper Río Paraguay is a large area of level plain with a variable geomorphology. This level plain has been and continues to be a site of alluvial deposition arriving from a variety of rivers, like the Paraguay and its tributaries that form the basin. Basically, the entire region is influenced by the following three factors. The first factor is the highly unstable hydrological system, influenced almost completely by the almost lack of slope of the terrain in the northeast-southwest direction; and an unstable hydrological regime of the rivers characterized by a temporary, sharp influx of water during summer. This results in a strong meandering plain determined by the geomorphology of the alluvial terrain. The normal, annual flood cycle determined by seasonal cycles of rainfall is the second factor, and the third consists of the heterogeneity of soils, including Quaternary soils of alluvial origin.

The Vegetation of the Littoral Zone of the Upper Río Paraguay

The physiognomy of the vegetative formations of the Upper Río Paraguay and the Pantanal de Nabileque, bordering the study area, is similar to that of the littoral vegetative formations of the Chaco. The flooded banks of the Río Negro, north of Bahía Negra, Paraguay, are dominated by a *vegetation mosaic* (Mereles and Aquino-Shuster, 1990). This *vegetation mosaic* is typical of the damper regions of the Chaco and consists of forests of *Schinopsis balansae* (Anacardiaceae), savanna-palm stands of *Copernicia alba* (Arecaceae), and flooded lowlands with a variety of aquatic-marsh vegetation dependent upon the geomorphology of the terrain and permeability of the soils. These conditions are essentially constant along the length of the Río Paraguay from the Río Negro to the border of the Provincia de Santa Fé, Argentina. *Vegetation mosaics* occupy practically the entire flood plains of the primary rivers of the Chaco basin, including: the Río Paraguay, the Río Pilcomayo (from northern Chaco, Paraguay to southern Chaco, Argentina), and the Río Bermejo (Argentinian Chaco).

As long as the forests of *Schinopsis balansae* and the palm stands of *Copernicia alba* are found over clay soils, which are structured and impermeable, they will likely be subject to periodic flooding. The time period of inundation is dependent upon two factors: the intensity of the rains and the swelling of the rivers, particularly the Río Paraguay, whose swelling is subject also to the quantity of rain falling in the Pantanal region to the north.

Forests flooded for 3–6 months of the year are dominated by *Schinopsis balansae*, also known as the *quebracho colorado*. Where this habitat converges with others in the region, like the xerophytic Chaco (Spichiger et al., 1995) and the Floresta Estacional Residual (Velos and Goes-Filho, 1982), other species become common. Examples of some of these other species are *Caesalpinia paraguariensis, Gleditsia amorphoides, Prosopis nigra, Enterolobium contortisiliquum, Lonchocarpus fluvialis* (Leguminosae), *Phyllostylon rhamnoides, Ziziphus mistol* (Rhamnaceae), *Chlorophora tinctoria* (Moraceae), *Trithrinax biflabellata, Syagrus romanzoffiana* (Arecaceae), *Diplokeleba floribunda, Magonia pubescens* (Sapindaceae), *Tabebuia heptaphylla, T. nodosa* (Bignoniaceae), *Astronium urundeuva, Schinus fasciculatus* (Anacardiaceae), *Maytenus ilicifolia* (Celastraceae), and *Crataeva tapia* (Capparaceae) among others (Mereles et al., 1992).

The palm stands of *Copernicia alba* occupy large areas within the mosaic of vegetation. The canopy is composed solely of *Copernicia alba*, except where the savanna-palm stands have been modified by human activity, such as raising cattle. This has occurred over large areas of the Chaco and along the littoral zone. In these cases, the palm stands have been invaded by very aggressive leñosas of the family Fabaceae (e.g., *Prosopis ruscifolia, Geoffroea decorticans*, and *Acacia caven*) (Mereles and Degen, 1997).

Under natural conditions, the palm stands of *Copernicia alba* are very open and one can observe additional species like *Mimosa hexandra* (Fabaceae), *Croton urucurana* (Euphorbiaceae), and *Thevetia bicornuta* (Apocynaceae). In lower areas with permanent water, large concentrations of herbaceous rizomatozas like *Cyperus giganteus, Scirpus californicus* (Cyperaceae), and *Typha domingensis* (Typhaceae) can be frequently observed. These permanent water conditions are more common near the outlet of the Río Nabileque into the Río Paraguay.

Palm stands can possess a very rich herbaceous layer that alternates between aquatic-marsh and terrestrial species. This results in high variability, and is due to the periodic flooding. Some of the more common species in this layer are *Aeschynomene sensitiva, A. falcata, Dischollobium pulchellum, Neptunia prostrata* (Leguminosae), *Byttneria filipes* (Sterculiaceae), *Commelina diffusa* (Commelinaceae), *Caperonia palustis* (Euphorbiaceae), *Eragrostis hypnoides, Leersia hexandra* (Poaceae), *Eupatorium candolleanum, Pterocaulum furfurescens* (Compositae), *Fimbristylis complanata, Eleocharis elegans, E. nodulosa* (Cyperaceae), *Heliotropium procumbens* (Boraginaceae), *Mayaca sellowiana* (Mayacaceae), *Ipomoea carnea* spp. *fistulosa* (Convolvulaceae), *Pavonia moronguii, Cienfuegosia drumondii* (Malvaceae), *Juncus densiflorus* (Juncaceae), and *Phyla reptans* (Verbenaceae) among others (Mereles et al., 1992).

In the third component of the *mosaic*, the flooded lowlands, one generally finds shelter for aquatic vegetation. These lowlands are often the remainders of old meanders of the Río Paraguay, its tributaries, or simply depressions in the terrain that remain flooded for most to all of the year. In many cases, species like *Pistia stratiotes, Eichhornia crassipes, Azolla* sp., or *Salvinia* sp. among others form large masses of floating vegetation.

Savannas of *Tabebuia aurea* are interspersed among the formations described above. These are usually over very permeable and less structured soils that are composed more of sand than clay (Chodat and Vischer, 1977). *Sapium subglandulosum, Cecropia pachystachya, Thevetia bicornuta*, and *Byrsonima intermedia* are species typically found growing in sandy soils, particularly if the area has been disturbed.

Finally, there is the shoreline vegetation. Common aquatic species can be divided into four different types: free floating, rooted on the shore, semi-submerged, and completely submerged. Each varies in function and type of aquatic habitat. The objective of this study was to determine the richness of the shoreline flora.

Methods

For the Río Paraguay, Río Napegue, and Río Aquidabán, only plants within the aqueous portion of the river-bed were sampled. Exceptions were at those sites where shoreline vegetation was partially covered by high-water, in which case these plants (including lianas and the epiphytes) were included. The diversity of habitats encountered along the length of the river in the area of study (habitat heterogeneity) was assessed. Representative plant specimens for each habitat were collected. Quantitative methods were not utilized. Flora along the shore was sampled by the terrestrial botany team (Beltrán, 2001).

Due to the paucity of aquatic vegetation in the Río Apa, the flooded banks were sampled with rapid transects between 500 and 800 meters inland from the shore. The vegetation further away was sampled by the terrestrial botany team. Specimens not collected on a transect, but found in the same area and considered informative were collected as well. Trees with a diameter at breast height greater than 10 centimeters were included in the count and collected if believed necessary, following the variable transect method (R.B. Foster, N.C. Hernández E., E.K. Kakudidi, and R.J. Burnham, unpublished manuscript).

Five duplicates for the majority of aquatic and marsh plants were collected. The first set of collections was deposited in the Herbario del Departamento de Botánica de la Facultad de Ciencias Químicas de la Universidad Nacional de Asunción (FCQ). The second set was deposited in The Field Museum in Chicago (FMNH). The third set was deposited in the Museo de Historia Natural de la Universidad Nacional Mayor de San Marcos, Lima (USM). Additional sets were placed in FCQ. A list of common names of some of the plants identified on the expedition are provided in Appendix 3.

Floral Diversity along the Río Paraguay

Four distinct habitat types were documented along the shoreline of the Río Paraguay, each of which is described below.

1) Semi-lotic habitat

This habitat is often encountered when rising waters have surpassed the height of the banks and overflowed the shoreline, inundating vegetative formations that are periodically covered by water (e.g., the savanna-palm stands of *Copernicia alba* (Arecaceae)). Under these conditions, there is the possibility of a very gentle current. The current velocity is normal along the shore, but varies inland depending upon the shoreline morphology. Where small cavities form, the water becomes almost completely dammed at the surface, but continues to move along the bottom. This main-

tains constant water clarity in the habitat, permitting the passage of light up to a meter in some places. Other areas are completely stagnant.

We assume that one would find semi-lotic conditions in areas where water has overflowed the banks due to seasonal flooding. These habitats are much more diverse than one would expect by chance, and probably account for the majority of aquatic floral diversity within the study area. This high diversity may be related to the dynamics of the waters in relation to their nutrients.

In a large number of cases, these flooded habitats are palm stands of *Copernicia alba*. In other cases, they are riparian forests. When comparing the two, the flooded palm stands possess a richer biodiversity.

Twelve localities considered semi-lotic were sampled (Appendix 4, Map 2). Twenty-nine free-floating species, six rooted semi-submergent species, five submergent species, one epiphytic species, and 21 marsh species (includes lianas, rooted semi-submergents, and others) were collected (Appendix 5).

2) Shoreline habitat

This habitat was relatively poor in aquatic vegetation and was fairly consistent among the sites sampled. The constant movement of the water currents along the shore limits the development of many species. Most plant species subsist by taking root, and may be anything from grasses to trees. There was very little floating vegetation.

Nine localities considered shoreline were sampled (Appendix 4, Map 2). Four free-floating species, 44 rooted semi-submergent species, and eight liana species were collected (Appendix 5).

3) Swamp habitat

This habitat is highly variable and very common along the entire length of the Upper Río Paraguay (Mereles, 1989). Some are more homogenous and almost without substrate (false soil), such as the "floating meadows" (large aggregations of floating vegetation). Others possess an appreciable amount of soil, facilitating the growth of shrubs and small trees. Species richness is proportional to the development and amount of soil (Mereles et al., 1992).

The area is teeming with examples of the first substrate-less stage ("floating meadows" in particular), probably due to the combination of strong winds and surface currents that break up the shoots of the species which form the base of the swamps. However, in places protected from the current and wind, such as streams and old meanderings presently isolated from the rivers, one is able to encounter more advanced stages containing considerable amounts of substrate. A larger variety of species are found at these sites.

Eichhornia crassipes and *E. azurea* (Pontederiaceae) (especially *E. crassipes*) constitute pioneering species for the swamps due to the tight joining of their shoots and ability to be dispersed with the aid of water currents.

In some swamps, arboreal vegetation is abundantly represented by *Aporosella chacoensis, Bauhinia bauhinioides,* and *Crataeva tapia.* These species can grow normally if the depth of substrate in the swamp is at least one meter.

Seven localities considered swamps were sampled (Appendix 4, Map 2). Eighteen free-floating species, ten lianas, two submergents, and 24 rooted semi-submergent species were collected (Appendix 5).

4)   Sand bank habitat

These are generally less rich in diversity, probably due to the poor quality of nutrients in their soils. However, the species encountered are very distinctive and characteristic of these sites. Some sites are large and accumulate water in the interior, forming small ponds with muddier soil that is enriched by the decomposition of organic material. Some species persist here, but are absent nearby where the soil is constantly leached by the water and contains only sand.

It is interesting to note that sand banks are scarce upstream from the San Alberto bank. This is probably due to the strength of the current and the width of the river.

Three localities considered sand banks were sampled (Appendix 4, Map 2). Two free floating species, 21 rooted semi-submergent species, and four lianas were collected (Appendix 5).

Floral Diversity in the Río Apa Sub-basin

Seven localities, each classified as a flooded bank, were sampled in the Río Apa (Appendix 4, Map 2). Twenty-eight aquatic plant species were collected on the flooded banks, and 22 species were collected along the shore, for a total of 50 species (Appendix 6). The floral diversity of the Río Apa sub-basin is much lower than the Upper and Lower Río Paraguay. Various factors are responsible for this, including the following:

• The Río Apa is a typical plains river, with a meandering course and depths of no more than one meter. Water levels can increase and decrease rapidly. Current velocity is significantly higher than the Río Paraguay, and two rapids (small cascades) are in the vicinity.

• The lack of habitat diversity was congruent with the scarcity of aquatic plants. For example, we rarely encountered marshes. The microhabitats present in the Río Paraguay were practically nonexistent in the Río Apa, being replaced by flooded banks where aquatic species rarely grow.

• Sand banks are very abundant. They appear to have been formed recently, hindering prolonged growth of vegetation.

• The presence of wooden debris like trunks and branches throughout the channel increases the current at some bends. This prevents the formation of appropriate habitats for the growth of plants. Tributaries supported a greater diversity of plants, but this did not carry into the Río Apa proper because the tributaries are of little volume and are unable to penetrate the Río Apa. This results in the formation of small ponds at the outlet of the tributaries. The water in these small ponds is stagnant and sufficiently turbid to prevent the growth of macrophytes.

In conclusion, the heterogeneity of habitats for the growth of aquatic plants is very low. The richer environments are found in some of the tributaries.

Recommended Conservation and Research Activities

• The highest diversity of aquatic plants is found in areas where water has overflowed the banks due to seasonal flooding. Seasonal fluctuations in water level play an important role in the local nutrient cycles and dispersal of species. Habitat modifications that disrupt seasonal flooding should be prevented.

• In the Río Apa sub-basin, tributaries supported a higher diversity of aquatic plants than the Río Apa proper. These small streams are easy to overlook, and are also easy to destroy. They should not be ignored when conservation decisions are made.

• Additional surveys of aquatic plants need to be conducted at different times during the year, particularly when the water levels are not as high as during this AquaRAP expedition.

Literature Cited

Adamoli, J. 1982. Vegetação do Pantanal. *In* Allem, A.A. and F.M. Valls (eds.). Recursos Forrageiros Nativos do Pantanal Mato-Grossense. Brasília: Empresa Brasileira de Pesquisa Agropecuária-EMBRAPA. Centro Nacional de Recursos Genéticos e Biotecnología e Centro de Pasquisa Agropecuária do Pantanal.

Cabrera, A. and A. Willink. 1976. Biogeografía de América Latina. Organización de Estados Americanos (O.E.A.), Serie Biología, 13:72–74.

Chodat, R. and W. Vischer. 1977. La végétation du Paraguay, (Historiae Naturalis Classica). *In* Cramer, J. (ed.)., Reimpresión de Edición. Vaduz. 588pp.

Mereles, F. 1989. Estudios de la vegetación dentro del mosaico bosque-sabana palmar- vegetación hidrófita en el Chaco boreal, Paraguay. Unpublished Ph.D. Thesis. Switzerland: Université de Genéve.

Mereles, F. and L. Aquino-Shuster. 1990. Los humedales en el Paraguay: breve reseña de su vegetación. La Revista Crítica, 3:49–66.

Mereles, F. and R. Degen. 1997. Leñosas colonizadoras e indicadoras de sitios modificados en el Chaco boreal, Paraguay. Rojasiana, 4:25–83.

Mereles, F., R. Degen, and N. Lopez de Kochalcka. 1992. Breve reseña de los humedales en el Paraguay. Amazoniana, 13:305–316.

Spichiger, R., R. Palese, A. Chautems, and L. Ramella. 1995. Origin, affinities and diversity hot spots of the paraguayan dendrofloras. Candollea, 50:517–537.

CHAPTER 4

Water Quality, Phytoplankton, and Benthic Invertebrates of the Upper and Lower Río Paraguay Basin, Paraguay

Francisco Antonio R. Barbosa, Marcos Callisto Faria Pereira, and Juliana de Abreu Vianna

Abstract

Water quality was assessed at 35 localities (Fig. 4.1). The waters of the Río Paraguay were generally slightly acidic (pH 6.0–6.5), with low oxygen levels (< 6.0 mg/L), low electrical conductivity (60–100 µS/cm), and temperatures ranging between 24–27° C. The waters of the Río Apa were neutral to slightly alkaline (pH 7.27–7.96), with variable levels of dissolved oxygen (4.2–7.13 mg/L), high electrical conductivity (163–376 µS/cm), and temperatures ranging between 22–29° C.

The phytoplankton communities were relatively diverse, and included representatives of several algae groups, including Chlorophyta, Euglenophyta, Chrysophyta, Bacillariophyta, and Cyanophyta. Benthic macro-invertebrates were collected at 33 sites. Chironomidae (Diptera) larvae and Oligochaeta were the dominant groups at 27 of the sites, representing 52% and 35% of the recorded organisms. Odonata, Trichoptera, Ephemeroptera, Chaoboridae, Ceratopogonidae, Corixidae, Conchostraca, Planaria, Nematoda, Hirudinea, Copepoda, Ostracoda, Bivalvia, and Gastropoda were collected less frequently.

Among the Chironomidae larvae, 26 taxa were identified. Some may be undescribed species. The diversity of benthic macro-invertebrates is high compared to other watersheds in South America. The majority of the genera are typical for herbaceous marshes, ponds, lakes, and slow moving portions of streams and rivers. Areas rich in decomposing vegetation exhibited a low diversity.

Overall, there was little evidence for direct contamination of the waters. The only exceptions were near Bahía Negra, where the burning and clearing of land could result in erosion and eutrophication.

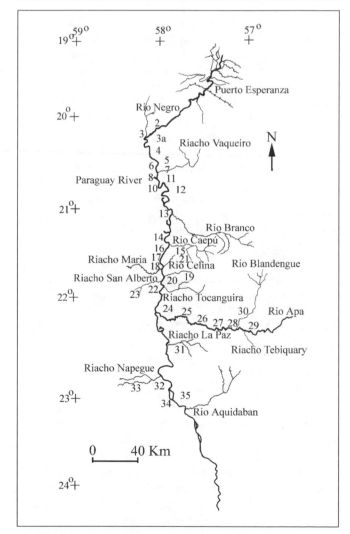

Figure 4.1. Map of the 35 stations sampled for water quality and benthic macro-invertebrates in the Upper and Lower Río Paraguay.

Introduction

The loss of biodiversity in aquatic ecosystems has received less attention than that in terrestrial ecosystems, despite knowledge of their physical, chemical, and biological degradation. The current state of aquatic habitats is clearly demonstrated by the increase in water-borne diseases (particularly in the tropics), decreasing fishery production, and diminishing water quality for human consumption, irrigation systems, and recreation. Furthermore, according to Master (1990), there are more aquatic organisms classified as rare to extinct (34% of fish, 75% of unionid mussels, and 65% of crayfish) than terrestrial ones (c. 11% to 14% of birds, mammals, and reptiles). Aquatic systems maintain a considerable biodiversity that is being lost mainly as a consequence of habitat deterioration.

The introduction of the watershed as the unit of study (Rigler and Peters, 1995) was of paramount importance in demonstrating that the implementation of conservation practices for terrestrial environments does not automatically include adequate protection of aquatic habitats (Junk, 1983; Barbosa, 1994). Despite this focus on watersheds, it is still unclear if the best conservation option is to consider the protection of total diversity of a given area or to focus on the highest possible number of rare species (see Pimm and Lawton, 1998). It is important to consider diversity in a dynamic context, including not only the preservation of the biotic elements but also the basic processes responsible for their maintenance (Norton and Ulanowicz, 1992). Furthermore, as suggested by Tundisi and Barbosa (1995), the adoption of the watershed as the conservation unit is fundamental since it integrates the multiple uses of natural resources and their surrounding social, economic, and cultural factors.

The primary objective of this paper is to summarize limnological data recorded during the 1997 AquaRAP expedition to the Upper and Lower Río Paraguay (20°09'7.9"S, 58°10'11.4"W) and to the sub-basin of the Río Apa (22°06'S, 57(55'W) (Fig. 4.1, Appendix 7). Small tributaries within these areas, namely the Arroyos Blandengue, Tebicuary, and Riacho La Paz, were also sampled. A total of 35 localities were visited during this period for which water temperature, pH, dissolved oxygen and electrical conductivity were measured *in situ*. Benthic macro-invertebrates were collected at 54 sampling stations.

The Upper Río Paraguay is characterized by the presence of several types of habitats among which floating meadows formed by several aquatic macrophytes, coastal areas dominated by flooded palms (*Copernicia alba*), and litoral vegetation dominated by *Schinopsis balansae* are the most common. The Lower Río Paraguay is dominated by floating meadows as well, but there were also vegetated sandy banks, small streams, and sandy/muddy beaches.

Material and Methods

Except for total alkalinity measurements and benthic organism identification, all variables were measured *in situ*. The samples for the physio-chemical measurements were collected at each site from below the surface (0.20 meter depth) with a plastic jar. The type of water, distance from the shore, depth, substrate type and vegetation were recorded. For the physio-chemical (water quality) characterization of the waters, the following variables were measured: water temperature (°C), electrical conductivity (μS/cm), pH, and dissolved oxygen (mg/L and % saturation). Total alkalinity was determined after titration with $0.1N$ H_2SO_4.

For the plankton samples, a 20 μm plankton net was used to filter 2 x 40 liters of sub-surface water at each site. Sediment samples were collected in triplicate using an Ekman-Birge dredge in such a way that 1 meter^2 was covered at each site. Samples were then rinsed through a 250 μm mesh and fixed in formalin. In the laboratory, the samples were washed through 1.0 and 0.50 millimeter mesh sieves, sorted under a stereomicroscope, and the organisms preserved in 70% ethanol. For identification, chironomid larvae were prepared using 10% lactophenol slides and their mouthparts examined under a 400x microscope.

Results and Discussion

Water Quality

The water in the Río Paraguay is generally slightly acidic (pH 6.0–6.5), with low oxygen levels (<6.0 mg/L), electrical conductivity ranging between 60–100 μS/cm, and temperatures ranging between 24–27°C (Table 4.1). No direct contamination could be observed at the sampling areas, thus we characterize the waters to be of good quality.

The water within the sub-basin of the Río Apa deserves special attention. It is generally neutral to slightly alkaline (pH 7.27–7.96), with temperatures varying between 22–28.9°C, variable levels of dissolved oxygen (4.2–7.13 mg/L), and high electrical conductivity (163–376 μS/cm) (Table 4.1).

This limestone area possesses waters of high quality, particularly at the headwaters that constitute water sources of paramount importance for the maintenance of the present aquatic biodiversity.

Plankton Communities

From a preliminary observation it can be stated that the phytoplankton communities were relatively diverse. Species of Chlorophyta, Euglenophyta, Chrysophyta, Bacillariophyta and Cyanophyta were recorded. The zooplankton is mainly dominated by Rotifera (*Lecane acus, Lecane flexilis, Lecane* sp., *Filinia sp., Conochilus* sp., *Polyarthra remata*), plus some Copepoda (mainly nauplius), Cladocera, and Protozoa. Detailed identification at the specific level is under way.

Table 4.1. Physico-chemical characteristics of the waters within the Upper/Lower Río Paraguay basin, September 1997. See Appendix 7 for site descriptions and locations.

Site	Air Temperature (°C)	Water Temperature (°C)	pH	Conductivity (µS/cm)	Dissolved Oxygen (mg/L)	Dissolved Oxygen (% sat)	Alkalinity (meqCO$_2$/L)
1	28.9	27.0	6.35	97.2	—	—	383.0
2	35.9	26.3	6.36	59.2	—	—	261.9
3	—	28.2	6.29	64.1	—	—	—
4	27.6	27.4	6.19	61.1	—	—	252.3
5	33.2	27.3	6.52	63.0	—	—	233.6
6	30.6	27.4	6.36	61.9	—	—	238.6
7	30.7	27.5	6.40	63.3	—	—	234.6
8	28.2	28.0	6.25	62.6	—	—	188.6
9	32.8	27.9	6.46	67.6	—	—	225.1
10	30.5	27.2	6.25	63.1	—	—	236.7
11	29.9	27.5	6.47	61.7	—	—	225.7
12	30.2	27.8	6.34	61.1	—	—	228.3
13	27.0	27.8	6.04	74.2	—	—	226.6
14	21.6	25.8	6.26	64.3	—	—	256.1
15	22.3	25.7	6.01	63.9	—	—	255.3
16	21.9	24.4	5.54	63.4	—	—	130.2
17	24.2	26.9	6.35	64.8	—	—	235.4
18	20.1	24.4	6.06	71.0	—	—	243.4
19	29.4	26.6	6.45	68.5	—	—	243.6
20	26.4	25.1	6.84	62.8	—	—	251.9
21	24.2	24.6	6.06	66.5	—	—	265.6
22	31.1	27.0	6.58	68.5	—	—	223.5
23	31.3	25.4	6.35	61.5	—	—	256.2
24	31.4	26.1	7.27	163.9	4.20	55.1	677.6
25	—	27.6	7.66	167.0	6.84	93.6	733.8
26	29.3	28.2	7.74	165.4	5.42	75.4	695.7
27	33.4	28.9	7.80	172.2	6.02	82.5	729.2
28	28.0	27.4	7.59	167.6	6.66	91.7	759.4
29	29.6	27.8	7.81	168.5	7.02	95.2	690.1
30	28.1	22.1	7.88	376.0	7.13	85.3	1790.6
31	29.0	24.3	7.96	375.0	6.30	80.9	1870.8
32	35.1	25.7	6.32	71.5	3.36	44.1	572.3
33	30.4	26.0	6.55	68.5	4.23	51.5	557.7
34	23.1	24.8	6.20	71.7	3.89	50.1	503.1
35	24.7	25.9	6.94	89.1	7.52	100.6	663.9

Table 4.2. Comparative information about taxa richness and number of chironomid (Diptera) genera in different watersheds in South America.

Taxa Richness	Number of Chironomidae genera	Watersheds in South America	References
16	26	Río Paraguay (present data)	Barbosa and Callisto (2000)
06	18	Upper Rio Doce (Serra do Cipó National Park, Brasil)	Galdean, Callisto, Barbosa, and Rocha (1999)
84	23	Río Piracicaba (Middle Doce River, Brasil)	Marques et al. (1999)
15	19	Headwaters of Rio São Francisco (Serra da Canastra National Park, Brasil)	Galdean, Callisto, and Barbosa (1999)
?	21	Río Paquequer (Rio de Janeiro State, Brasil)	Nessimian and Sanseverino (1998) Sanseverino and Nessimian (1998)
29	31	Río Negro (Patagonia, Argentina)	Wais (1990)
14	17	Río Gravataí (Rio Grande do Sul, Brasil)	Bendati et al. (1998)
80	35	Upper Río Paraná (Brasil)	Takeda et al. (1997)
11	15	Río Trombetas (Brazilian Amazonia)	Callisto et al. (1998)

Benthic Macro-invertebrates

A total of 2,213 benthic macro-invertebrates specimens were recorded from 33 sampling stations (Appendix 8). Chironomidae larvae and Oligochaeta were the dominant groups in 27 stations, representing respectively 52% and 35% of the recorded organisms. Other groups collected in smaller numbers include Odonata, Trichoptera (*Oxyethira*), Ephemeroptera, Chaoboridae, Ceratopogonidae, Corixidae, Conchostraca, Planaria, Nematoda, Hirudinea, Copepoda, Ostracoda, Bivalvia, and Mollusca (*Pomacea*) (Appendix 8). Among the Chironomidae larvae, 26 taxa were identified. The most common (listed from most to least abundant) were *Nimbocera paulensis, Polypedilum, Chironomus, Ablabesmyia, Goeldichironomus, Fissimentum desiccatum, Harnischia, Nilothauma, Parachironomus, Stenochironomus, Asheum, Coelotanypus,* and *Djalmabatista*. At lower densities there were recorded larvae of *Axarus, Beardius, Cladopelma, Cryptochironomus, Phaenopsectra, Denopelopia, Labrundinia, Macropelopia, Tanypus, Corynoneura, Cricotopus, Nanocladius* and *Thienemanniella*, along with specimens belonging to the many yet undescribed, endemic species of South America (see Fittkau and Reiss, 1973).

The diversity of benthic macro-invertebrates can be considered high when compared with other watersheds in South America (Table 4.2). Furthermore, the majority of the recorded genera are typical of herbaceous marshes, ponds, lakes, and the slower moving portions of streams and rivers. They apparently prefer soft waters of low alkalinity and slightly acidic to circumneutral pH.

On the other hand, areas rich in decomposing vegetation exhibited a low diversity of benthic organisms. Sampling stations 2 and 3 (west bank of the Río Paraguay, rich in aquatic macrophytes) and station 23 (Riacho San Alberto, flooded area covered mainly with *Copernicia alba*) are good examples, with each dominated (80-100%) by *Chironomus* larvae. These larvae prefer habitats rich in decomposing organic matter, low oxygen concentrations, and oftentimes waters of low quality.

Some of the recorded genera usually feed on macrophyte detritus, and others are characteristically miners of decomposing plant tissues (e.g., *Goeldichironomus, Polypedilum*, and *Stenochironomus*). Finally, some of these organisms typically live on periphyton (*Beardius, Asheum, Cryptochironomus*) or feed directly on filamentous algae (*Cricotopus*).

Two of the recorded species deserve special attention: *Nimbocera paulensis* Trivinho-Strixino and Strixino (1991) and *Fissimentum desiccatum* Cranston and Nolte (1996). *Nimbocera paulensis* was first described as typical of rheolimnic conditions, shallow environments with sandy and organic sediments. *Fissimentum desiccatum* is a typical inhabitant of the potamal region, also areas with sandy and organic sediments but possessing decomposing aquatic macrophytes. According to Cranston and Nolte (1996), the larvae of *F. desiccatum* build galleries in the sediment that allow them to resist dry periods and return when the sediment is rehydrated. At sampling stations 24 and 35, intense predation by the carnivorous larvae of *Coelotanypus* on larvae of *F. desiccatum, N. paulensis* and *Phaenopsectra* was observed.

Cladopelma, Axarus, Harnischia, Nilothauma, Parachironomus, Corynoneura, Nanocladius and *Thienemanniella* have wide distributions, and were collected from all the existing microhabitats. The observed benthic macro-invertebrate communities possessed a high diversity and abundance of carnivorous larvae of the genera *Ablabesmyia, Coelotanypus, Labrundinia, Denopelopia, Macropelopia, Djalmabatista* and *Tanypus*. It is suggested that these carnivorous larvae are the major controlling agent of the richness of other chironomids, thus resulting in a rather high diversity in comparison with other aquatic systems in South America (e.g., Nolte, 1989; Strixino and Trivinho-Strixino, 1991; Higuti et al., 1993; Nessimian, 1995; Nessimian and Sanseverino, 1995; Callisto et al., 1998).

A tentative classification of the studied areas based upon benthic macro-invertebrate diversity

Although the present results were obtained from a single survey (at the end of the rainy season), they provide a first assessment of water quality as well as the structure and distribution of the benthic macro-invertebrate communities for the area. Furthermore, the study also allowed us to identify eleven types of microhabitats. They are silt and clay, silt, fine sand, silt with mosses, silt with filamentous algae, silt with macrophyte detritus, coarse sand with plant debris, fine particulate organic matter, coarse particulate organic matter, aquatic macrophytes, and filamentous algae. It must be pointed out that the sampling period was rather atypical, with higher water levels than usual for this time of the year.

Based on the diversity index (H; Shannon-Wiener) of Chironomidae larvae and the recorded substrate types (Appendix 9), Barbosa and Callisto (2000) suggested three major divisions along the Upper/Lower Río Paraguay in relation to their benthic macro-invertebrate community diversity. This information should be considered when making conservation decisions.

(a) Areas of **high diversity:** with an H' > 3.0 and represented in this survey by sampling stations 30 and 31, located respectively in the streams Blandengue and La Paz. Here the highest number of microhabitats was found including, sand, gravel, periphyton, aquatic mosses, and a stand of *Cabomba* sp. These are relatively well-protected areas with riparian vegetation and no significant human influences recorded.

(b) Areas of **intermediate diversity:** $1.8 \leq H' \leq 3.0$, represented by sampling stations 2, 3A, 5, 6, 10, 15, 18, 26, 27, 28, and 29, which are flooded areas or areas rich in aquatic macrophytes with sediments rich in organic matter in distinct decomposition stages, receiving both allocthonous (riparian vegetation) and autocthonous (aquatic macrophytes, mosses and filamentous algae) contributions.

(c) Areas of **low diversity:** with an H' < 1.8 and represented by sampling stations 3, 4, 7, 8, 12, 14, 17, 21, 22, 23, 24, 25, 31A, and 35. The sediments found here are mainly formed by silt and clay, but silt and filamentous algae, silt and aquatic mosses, fine sand and organic matter in deep decomposition stage also occur. These areas show low diversity of microhabitats.

Recommended Conservation and Research Activities

- Overall, the Río Paraguay basin is relatively well preserved, with the exception of the northern part near Bahía Negra. This area is primarily threatened as a consequence of disorganized land use resulting in widespread deforestation in order to form pasture land. Erosion is a direct consequence, thus affecting the water and its living communities. Furthermore, the burning of natural vegetation, particularly large areas of *Copernicia alba* (palm trees), constitutes another considerable threat that may result in eutrophication of these waters. Habitat impact studies need to be conducted to determine how to manage the land properly.

- Limestone areas possess waters of high quality. Steps need to be taken to preserve these sites.

- The diversity of macro-invertebrates can be considered high when compared with other watersheds in South America. Most of these species are found in habitats created by seasonal flooding. Habitat modifications that disrupt the annual changes in water level should be prevented. An example of habitat modification is the Hidrovia Paraguay-Paraná, which will affect directly the seasonal flooding regime of the Río Paraguay and indirectly other streams within the watershed, resulting in the loss of certain habitats and their communities.

- The Río Verde is under severe threat of salinization.

Literature Cited

Barbosa, F.A.R. (ed.). 1994. Workshop: Brazilian programme of conservation and management of inland waters. Acta Limnologica Brasiliensia, Belo Horizonte. 222pp.

Barbosa, F.A.R. and M. Callisto. 2000. Rapid assessment of water quality and diversity of benthic macro-invertebrates in the Upper and Middle Río Paraguay using the Aqua-RAP approach. Verhandlungen der Internationalen Vereinigung für Theoretische und Angewandte Limnologie 27: (in press).

Bendati, M.M.A., C.R.M. Maizonave, E.D. Olabarriaga, and R.M. Rosado. 1998. Use of the benthic macroinvertabrate community as a pollution indicator in the Gravataí River (RS, Brazil). Verhandlugen der Internationalen Vereinigung für Theoretische und Angewandte Limnologie, 26:2019–2023.

Callisto, M., F.A. Esteves, J.F. Gonçalves Jr., and J.J.F. Leal. 1998. Impact of bauxite tailings on the distribution of benthic macrofauna in a small river (`igarapé`) in central Amazonia, Brazil. Journal of Kansas Entomological Society, 71:447–455.

Cranston, P.S. and U. Nolte. 1996. *Fissimentum,* a new genus of drought-tolerant Chironomini (Diptera: Chironomidae) from the Americas and Australia. Entomological News 107:1–15.

Fittkau, E.J. and J. Reiss. 1973. Amazonische Tanytarsini (Chironomidae, Diptera) I. die riopreto-gruppe der gattung *Tanytarsus.* Studies on the Neotropical Fauna, 8:1–16.

Galdean, N., M. Callisto, and F.A.R. Barbosa. 1999. Benthic macroinvertebrates of the head-waters of river São Francisco (National Park of Serra da Canastra, Brazil). Travaux Museum Natl. Hist. Nat. Grigore Antipa, vol. XLI: 455–464

Galdean, N., M. Callisto, F.A.R. Barbosa, and L.A. Rocha. 2000. Lotic ecosystems of Serra do Cipó, southeast Brazil: water quality and a tentative classification based on the benthic macroinvertebrate community. Aquatic Ecosystem Health and Management, 3:545–552.

Higuti, J., A.M. Takeda, and A.C. Paggi. 1993. Distribuição espacial das larvas de Chironomidae (Insecta, Diptera) do Rio Baía (MS-Brasil). Revista UNIMAR, 15(supplement):65–81.

Junk, W.J. 1983. Ecology of swamps on the middle Amazon. *In* Gore, A.J.P. (ed.). Mires: swamps, bog, fen and moor. Amsterdam: Elsevier Scientific Publishing Company. Pp. 269–294.

Marques, M.M.G.S.M., F.A.R. Barbosa, and M. Callisto. 1999. Distribution and abundance of Chinomidae (Diptera, Insecta) in an impacted watershed in south-east Brazil. Revista Brasileira de Biologia, 59:1–9.

Master, L. 1990. The imperiled status of North American aquatic animals. Biodiversity Network News (Nature Conservancy), 3:1-2,7-8.

Nessimian, J.L. 1995. Composição da fauna de invertebrados bentônicos em um brejo entre dunas no litoral do Estado do Rio de Janeiro, Brasil. Acta Limnologica Brasiliensia, 7:41–59.

Nessimian, J.L. and A.M. Sanseverino. 1995. Structure and dynamics of chironomid fauna from a sand dune marsh in Rio de Janeiro State, Brazil. Studies on Neotropical Fauna and Environment, 30:207–219.

Nessimian, J.L. and A. Sanseverino. 1998. Trophic functional characterization of chironomidae larvae (Diptera: Chironomidae) in a first order stream at the mountain region of Rio de Janeiro State, Brazil. Verhandlungen der Internationalen Vereinigung für Theoretische und Angewandte Limnologie, 26:2115–2119.

Nolte, U. 1989. Observations on neotropical rainpools (Bolivia) with emphasis on Chironomidae (Diptera). Studies on Neotropical Fauna and Environment, 24:105–120.

Norton, B.G. and R.E. Ulanowicz. 1992. Scale and biodiversity policy: a hierarchical approach. Ambio, 21:244–249.

Pimm, S.L. and J.H. Lawton. 1998. Planning for biodiversity. Science 279:2068–2069.

Rigler, H.F. and R.H. Peters. 1995. Science and limnology: excellence in ecology. New York: Ecology Institute. 245pp.

Sanseverino, A. and J.L. Nessimian. 1998. Habitat preferences of Chironomidae larvae in an upland stream of Atlantic Forest, Rio de Janeiro State, Brazil. Verhandlungen der Internationalen Vereinigung für Theoretische und Angewandte Limnologie, 26:2141–2144.

Strixino, G. and S. Trivinho-Strixino. 1991. Chironomidae (Diptera) associados a sedimentos de reservatórios: significado dos diferentes povoamentos. An. Sem. Reg. Ecol., 6:151–168.

Takeda, A., G.Y. Shimizu, and J. Higuti. 1997. Variações espaço-temporais da comunidade zoobêntica. *In* Vazzoler, A.E.A.M., A.A. Agostinho, and N.S. Hahn (eds.). A planície de inundação do alto rio Paraná: aspectos físicos, biológicos e socioeconômicos. Ed. Universidade Estadual de Margingá. Pp. 158–177.

Trivinho-Strixino, S. and G. Strixino. 1991. Duas novas espécies de Nimbocera Reiss (Diptera, Chironomidae) do Estado de São Paulo, Brasil. Revista Brasileira de Entomologia, 35:173–178.

Tundisi, J.G. and F.A.R. Barbosa. 1995. Conservation of aquatic ecosystems: present status and perspectives. *In* Tundisi, J.G., C.E.M. Bicudo, and T. Matsumura-Tundisi (eds.). Limnology in Brazil. Rio de Janeiro: BAS, BIS.

Wais, I.R. 1990. A checklist of benthic macroinvertabrates of the Negro River basin, Patagonia, Argentina, including an approach to their functional feeding groups. Acta Limnologica Brasiliensia, 3:829–845.

CHAPTER 5

Diversity, Distribution, and Habitats of the Macro-invertebrate Fauna of the Río Paraguay and Río Apa, Paraguay, with Emphasis on Decapod Crustaceans

Célio Magalhães

Abstract

The aquatic macro-invertebrate fauna was surveyed at 54 collecting stations along the Upper Río Paraguay, between the mouth of the Río Negro and Río Aquidabán, and along the middle section of the Río Apa. The collecting efforts were directed towards an assessment of the decapod crustacean fauna, but other groups were incidentally collected. Sampled habitats included river beaches, backwaters, lagoons, rapids, flooded plains, floating vegetation, submerged litter, and cryptic habitats (such as in submerged logs, among dead leaves, and between and under stones in the bottom of streams). Collections were qualitative only and performed with seine nets, dip nets, traps, and by hand.

Thirteen species of decapod crustaceans were found: six species of Palaemonidae and one of Sergestidae shrimps, and six species of Trichodactylidae crabs. The occurrence of *Pseudopalaemon* sp. is the first record of this genus in the Río Paraguay basin, and is probably a new species. The shrimp *Macrobrachium amazonicum* and the crabs *Trichodactylus borellianus* and *Sylviocarcinus australis* were the most frequent species. The aquatic insect fauna was the richest in number of species, with 83 species in 34 families and ten orders. Coleoptera, Hemiptera, and Odonata were most often collected. Molluscs were an important group in regards to abundance and biomass, with at least seven species of Gastropoda and three species of Bivalvia. Other less representative groups included periphytic and parasitic crustaceans (Conchostraca, Ostracoda, Amphipoda, Isopoda, and Branchiura), annelids, and platyhelminth worms.

Introduction

The macro-invertebrate collecting effort of this AquaRAP expedition was directed mainly to assess the decapod crus-
tacean fauna. Other groups, such as molluscs, insects, periphytic and parasitic crustaceans, annelid and plathyhelminth worms, were also collected, but sampling was either incomplete or superficial, and therefore a detailed evaluation of that fauna was not possible. The results and a discussion are presented below, with emphasis on the decapod crustaceans.

The freshwater shrimp and crab fauna of Paraguay is not very well-studied, although sporadic citations indicate the presence of most of the species that occur in the entire Paraguay and Lower Paraná river basins. The main original records of this group in Paraguay were made by Stimpson (1861), Nobili (1896), Rathbun (1906), Hansen (1919), Bott (1969), Rodríguez (1992), and Magalhães and Türkay (1996b). Kochalka (1996) listed the species present in the National Museum of Natural History of Paraguay, but did not provide information on the locations where the specimens were collected. Based on these citations, six species of crabs and six species of shrimps are known for the country. Even so, information about the distribution of these species in Paraguayan territory and the main habitats occupied by them are still scattered and imprecise. Therefore, the result of the AquaRAP expedition represents an important contribution to the knowledge of the fauna in a poorly surveyed area of the country.

Methods

Sampling was done mainly during the day time, using a dip net, a seine net (mesh size: 5 millimeters), and traps. Night sampling, using a flash light, was performed to capture crabs along the shoreline. Specimens were fixed and preserved in 70% ethyl alcohol. Independent of the collecting methods and the sampling period, the samples were bulked according to the sub-basins considered in Appendix 10 and according to the habitats specified in Appendix 11. The Cerritos Pão de

Açúcar area was used to divide the surveyed stretch of the Río Paraguay at about 21° 26'S, 57° 55' W into the Upper and Lower Paraguay sub-basins. The Upper Paraguay sub-basin included the area of the mouth of the Río Negro, because it did not present significant differences either in specific composition or in habitat types in relation to the Río Paraguay. In this sub-basin, the river presented wide floodplain areas with large extensions of camalotes (meadows of floating and rooted aquatic macrophytes), as well as many channels and secondary branches. In the Lower Paraguay sub-basin, south of the cerritos area, the river is restricted to its banks and has smaller areas of floodplains and isolated lagoons and tributaries. The Río Apa sub-basin comprises the Río Apa and its tributary streams (or arroyos), characterized by defined banks, riparian forests, sandy beaches, and areas with entanglements of submerged logs and dead leaves. The Riacho La Paz sub-basin was considered only by sampling done in its headwater area, which is characterized by a strong current, a rocky bed, and submerged litter. Sampling performed at the mouth of the Riacho La Paz was considered as part of the Lower Paraguay sub-basin due to the similarities between the two areas.

The identification of the specimens was based on the descriptions of Holthuis (1952), Omori (1975), Rodríguez (1992) and Magalhães and Türkay (1996a,b), and on comparisons with specimens from the Crustacean Collection of the Instituto Nacional de Pesquisas da Amazônia, in Manaus, Brasil.

Crustacea: Decapoda

A total of 13 species of decapods from three families and 10 genera were found (Appendix 10). Most of the species have wide distributions, also occurring in other South American river basins, especially in the Amazon basin. The number of species found during this survey is in accordance with the specific composition expected for the area. The only remarkable novelty is the occurrence of *Pseudopalaemon* sp., which represents the first record of the genus for the Río Paraguay basin. The species is probably new, since it differs both from the Amazon species of the genus and from *Pseudopalaemon bouvieri*, recorded for southern Brasil, Uruguay, and Argentina. Unfortunately, just three specimens were collected and this is an insufficient amount for a conclusive definition of their specific identity.

Of the species collected in this survey, the crab *Sylviocarcinus australis* and the shrimp *Pseudopalaemon* sp. are the only species that do not occur in other river basins. The endemicity of *Pseudopalaemon* sp. would be more restricted than that of *S. australis*, since the former has only been collected in the Río Apa to date, while the latter shows a wider distribution throughout the Río Paraguay and part of the Lower Río Paraná (Magalhães and Türkay, 1996b). During

this assessment, there were no species conspicuously absent from the area, based on our expectations. Two trichodactylid crabs previously reported for the Río Paraguay basin did not appear in the samples, but this is understandable. *Trichodactylus parvus* seems to be endemic to the northern Pantanal, in the Mato Grosso state, Brasil (Moreira, 1912). *Goyazana castelnaui*, though having a wide distribution, is a species typical of high plain rivers and is present only in the headwaters of the northern Pantanal (Magalhães and Türkay, 1996c). Another trichodactylid crab reported for Paraguay, *Trichodactylus kensleyi*, was also not found. The distribution of this species is seemingly restricted to the east part of the country, in the Upper Río Paraná basin (Rodríguez, 1992).

The decapod fauna of the surveyed area of the Río Paraguay shows some similarities with that of the Lower Río Paraná (downstream from the Itaipu Dam, near Foz do Iguaçú, Brasil, and Ciudad de Leste, Paraguay). Representatives of Aeglidae and Parastacidae were not found, but the distribution of these groups does not include the Middle and Upper Río Paraguay basin (Manning and Hobbs, 1977; Bond-Buckup and Buckup, 1994). However, considering only the three families recorded (Sergestidae, Palaemonidae and Trichodactylidae), almost 50% of the species collected in the surveyed area also occur in the Lower Paraná basin. The trichodactylid crabs have five species common to both systems, one species (*Zilchiopsis oronensis*) with a northern distribution (Ríos Paraguay and Amazon basins) and two species (*Zilchiopsis collustinensis* and *Trichodactylus kensleyi*) occurring in the Río Paraná basin (Table 5.1). Only three species of shrimps occur in both basins (*Macrobrachium amazonicum*, *M. borellii* and *Acetes paraguayensis*). It seems that the distribution of the remaining four species do not extend more southward than the area surveyed. Two species from the Lower Río Paraná (*Palaemonetes argentinus* and *Pseudopalaemon bouvieri*) are not known from the Río Paraguay (Table 5.1).

The decapod fauna of Río Paraguay is less diverse than that of the Amazon basin. While a total of 15 species is known for the former basin, approximately 53 species are recorded from the latter. However, most of the components of the fauna collected during this AquaRAP expedition (10 species) are also present in the Amazon basin. These two basins have large floodplain areas with similar patterns of aquatic vegetation and annual regimes of inundation. Several of the shrimp and crabs species occurring in the floodplains of the Amazon are also present in the same kinds of environments within the Río Paraguay basin.

A more precise idea of the distribution of the species in the studied area might have been obtained if not for the effects of the high water level during the collecting trip. In such a situation, delimitation and access to the habitats commonly occupied by these crustaceans is more difficult. Also, a more careful survey was restricted by the short sampling time in each explored site—a situation particularly critical in the

Table 5.1. Comparative list of the decapod crustacean species (Palaemonidae, Sergestidae, Trichodactylidae) collected in the Upper Río Paraguay by the AquaRAP expedition (September, 1997) with those reported for the Lower Río Paraná basin. Distribution data from Stimpson (1861), Nobili (1896), Rathbun (1906), Hansen (1919), Bott (1969), Rodríguez (1992), and Magalhães and Türkay (1996b).

Taxon	Río Paraguay	Lower Río Paraná
Palaemonidae		
Macrobrachium amazonicum	■	■
Macrobrachium borellii	■	■
Macrobrachium brasiliense	■	—
Macrobrachium jelskii	■	—
Palaemonetes ivonicus	■	—
Palaemonetes argentinus	—	■
Pseudopalaemon bouvieri	—	■
Pseudopalaemon sp.	■	—
Sergestidae		
Acetes paraguayensis	■	■
Trichodactylidae		
Dilocarcinus pagei	■	■
Poppiana argentiniana	■	■
Sylviocarcinus australis	■	■
Trichodactylus borellianus	■	■
Trichodactylus kensleyi	—	■
Valdivia camerani	■	■
Zilchiopsis collastinensis	—	■
Zilchiopsis oronensis	■	—

Riacho La Paz headwaters, which were limited to a single, short, daytime sampling.

The specific composition with regards to the sub-basins (Appendix 10) shows that *Trichodactylus borellianus, Sylviocarcinus australis* and *Macrobrachium amazonicum* were the most frequent species, the former being the only species present in all the sub-basins. Appendix 10 indicates a larger diversity of shrimps in the Lower Río Paraguay. In this area, only two species were not present: *Macrobrachium brasiliense* (probably due to the absence or impossibility of sampling its preferential habitat) and *Pseudopalaemon* sp., restricted to the Río Apa sub-basin. In the Upper Río Paraguay sub-basin, only *M. amazonicum* was found. There was no significant difference among the Upper and Lower Río Paraguay sub-basins with respect to crabs. *Valdivia camarani* and *Poppiana argentiniana* did not appear in the samples of the Lower Río Paraguay, but these are apparently not very frequent species and their absence should be related to the sampling restrictions mentioned above. These restrictions could also have caused of the absence of the shrimps *Macrobrachium jelskii* and *Palaemonetes ivonicus* in the Upper Paraguay sub-basin samples.

The lower species richness in the Río Apa and Riacho La Paz sub-basins could be explained by the characteristics of those aquatic systems. Basically, they do not have the large floodplain areas and the extensive meadows of aquatic vegetation. The decapod fauna present is mainly that which occupies cryptic habitats, and the Riacho La Paz was explored only very briefly, in limited headwaters area. Even so, it is unlikely that species other then *T. borellianus, M. brasiliense*, and, perhaps, *S. australis*, occur there, because the others are not typical of such fast flowing, rocky bed environments.

The three more abundant species in the collections were *Macrobrachium amazonicum, Trichodactylus borellianus*, and *Dilocarcinus pagei*. These species occurred in a great variety of habitats, being particularly common among the roots of the floating vegetation that forms the large extensions of macrophytes banks (camalotes), especially *Eichhornia crassipes, E. azurea*, and *Paspalum stoloniferum*. The shrimps were present either in the main channel of the Río Paraguay or in its several lateral branches, both among the macrophytes banks and along the sandy-muddy beaches. *Macrobrachium amazonicum* and *D. pagei* were also found in lagoons and in flooded forest areas. *Trichodactylus borellianus* was also found in fast flowing areas and in cryptic habitats, such as in submerged logs, among dead leaves, and between and under stones in the bottom of streams, and also in terrestrial environments, under stones or other objects close to the shoreline. *Macrobrachium amazonicum* also occurred in the Río Apa. However, its abundance was much lower than that of the Río Paraguay due to the absence of floating vegetation. In the former sub-basin, it was found in sandy-muddy beaches and among submerged litter.

Most of the decapod species, especially the crabs, were found along the shoreline of the Río Paraguay, either in gentle cliff banks or on sandy-muddy beaches (Appendix 11). Some, however, had a more restricted occurrence. The shrimp *Palaemonetes ivonicus* was found in small numbers associated with the floating vegetation in an isolated lagoon. A species commonly found among the roots of aquatic macrophytes, its absence in environments such as the main river could perhaps be due to seasonal factors. These factors could also explain the occasional occurrence of *M. jelskii* and *M. borellii. Acetes paraguayensis* is typically a pelagic shrimp, and its occurrence in the channel of the Río Paraguay was foreseeable; however, the capture of some specimens in isolated lagoons was occasional. The predominance of the shrimp *M. brasiliense* in the Río Apa and Riacho La Paz sub-basins, and its probable absence in the Río Paraguay habitats, can be understood by its preference for cryptic habitats found more typically in the rivers and fast flowing streams bordered by the riparian forest of those sub-basins. The crabs occurred in marginal areas (banks or beaches), associated or not with the aquatic vegetation. In such habitats, the crab *Sylviocarcinus australis* was found in both the Río Paraguay sub-basins and the Río Apa sub-basin. Its absence from the Riacho La Paz sub-basin is probably due to sampling restrictions. The samples of *Pseudopalemon* sp., *Poppiana argentiniana,* and *Valdivia camerani* were too low to allow any inference concerning their preferential habitats.

The shrimp *Macrobrachium amazonicum* was the most abundant species in the collections, occurring in several habitats. A species that includes a larval development with several free-swimming planktotrophic zoea larvae, its proliferation is related to a reasonable planktonic-rich environment and considerable aquatic macrophyte growth, essential for the feeding and the protection of the species in that phase of its life cycle (Magalhães and Walker, 1988; Collart Odinetz, 1991). The wide distribution and abundance of *M. amazonicum* in the Río Paraguay sub-basins reflects the wealth of this aquatic system. In the Río Apa sub-basin, lacking in floating macrophytes, its occurrence suggests that even this system presents a reasonable productivity, capable of sustaining a significant population of this shrimp.

It is known that shrimps and crabs make up a part of the diet of several groups of fishes, including the large catfishes (Siluriformes) (Goulding and Ferreira, 1984), some of which are valued for game fishing. It was verified that indigenous populations use this resource, capturing the crabs to sell as bait for game fishermen. These crustaceans are also prey of caimans, turtles, mammals, and aquatic birds, which suggests that they play an important role in the food web of this aquatic ecosystem.

Other Macro-invertebrates

Although the sampling efforts were mainly directed toward decapod assessment, other macro-invertebrate groups appeared incidentally in the collections (Appendix 10). Among the crustaceans, specimens of Conchostraca, Ostracoda, and Amphipoda (Hyallelidae: *Hyallela pernix*) were found associated with the aquatic vegetation, mainly *Eichhornia azurea* and *E. crassipes,* making up part of the periphytic community. As such, they were only captured in the Río Paraguay sub-basins and not in the Río Apa and Riacho La Paz sub-basins, where such habitats were not available. Two species of a crustacean fish ectoparasite (Branchiura: *Dolops* spp.) were found on "dorado" (Characidae: *Salminus maxillosus*) and on "piranha" (Serrasalmidae: *Pygocentrus nattereri*).

The molluscs were one of the main macro-invertebrate groups collected, both in terms of abundance and biomass. At least seven species (five families) of gastropods and three species (two families) of bivalves were found. The most frequent species were the gastropods *Pomacea canaliculata* (Ampullariidae) and *Biomphalaria* sp. (Planorbidae). The former only occurred in the Río Paraguay sub-basins, while the latter was also found in the Río Apa sub-basin. The

bivalves of the family Mycetopodidae were more frequent in the Río Apa sub-basin, although at least one species, *Anodontites crispatus tenebricosus,* occurred in the Lower Río Paraguay sub-basin.

The floating and rooted aquatic vegetation represented the main habitat exploited by the molluscs in the Río Paraguay sub-basins. They were commonly found in the meadows of floating macrophytes, and were also present in the vegetation banks along the shoreline, in the swamps, and in the flooded forest areas. In the Río Apa sub-basin, they were found along marginal areas, in the submerged litter, and even in cryptic habitats, such as holes in submerged logs.

The aquatic insect fauna was the richest in number of species. In all, 83 species in 34 families and ten orders were found, among which Coleoptera, Hemiptera, and Odonata were the most representative. Inferences on the preferential habitats and the distribution areas were affected by the irregular sampling efforts. In spite of this, a larger species richness in the Upper Río Paraguay sub-basin was noted, probably due to the largest extensions of aquatic vegetation meadows occurring in that area. Some of the species listed in Appendix 10 are not considered aquatic and they were collected by chance. This is the case of some Coleoptera groups that can live on rock substrates close to the shoreline (such as some Carabidae) or associated with the emerged portion of the macrophytes (such as Lampyridae, Chrysomellidae, and Curculionidae). Also, the capture of some Scarabaeidae, whose representatives are saprophagous, could have been occasional.

Recommended Conservation and Research Activities

- Decapod and macro-invertebrate diversity in the Río Paraguay was highest in habitats with floating and rooted aquatic vegetation. These habitats are dependent upon seasonal floods. Habitat modifications that disrupt the annual changes in water levels should be prevented.

- Habitat heterogeneity in the Río Apa is low. Most of the decapod and macro-invertebrate diversity is associated with cryptic habitats (i.e., submerged logs), which provide opportunities for cover and foraging. These habitats need to be conserved.

- Shrimps and crabs are a food source for some game fishes. Indigenous populations capture crabs and sell them as bait to fishermen. At the moment, capture of specimens is not believed to represent a risk to the natural reserve of this species in the environment, but small-scale breeding of this species could be developed into additional income for the local population.

- Crabs and shrimps are relatively easy to breed in captivity. If an environmental disaster were imminent, then specimens could be collected and maintained as "recolonization stocks" for a later date. This would help to preserve the uniqueness of the Paraguayan populations.

Literature Cited

Bond-Buckup, G. and L. Buckup. 1994. A família Aeglidae (Crustacea, Decapoda, Anomura). Arquivo Zoologia, São Paulo, 32:159–346.

Bott, R. 1969. Die süsswasserkrabben Süd-Amerikas und ihre stammesgeschichte. Eine revision der Trichodactylidae und der Pseudothelphusidae östlich der Anden (Crustacea, Decapoda). Abhandlungen der Senckenbergischen Naturforschenden Gesellschaft, 518:1–94.

Collart Odinetz, O. 1991. Strategie de reproduction de *Macrobrachium amazonicum* en Amazonie centrale (Decapoda, Caridea, Palaemonidae). Crustaceana, 61:253–270.

Goulding, M. and E. Ferreira. 1984. Shrimp-eating fishes and a case of prey switching in Amazon rivers. Revista Brasileira de Zoologia, 2:85–97.

Hansen, H.J. 1919. The Sergestidae of the Siboga Expedition. Siboga Expedition, 38:1–65.

Holthuis, L.B. 1952. A general revision of the Palaemonidae (Crustacea Decapoda Natantia) of the Americas. II. The subfamily Palaemonidae. Occasional Papers, Allan Hancock Foundation, 12:1–396.

Kochalka, J.A. 1996. Lista de invertebrados de Paraguay pertenecientes a las colecciones del Museo Nacional de Historia Natural del Paraguay: Clase Crustacea. *In* Romero Martinez, O. (ed.). Colecciones de Flora y Fauna del Museo Nacional de Historia Natural del Paraguay. Museo Nacional de Historia Natural del Paraguay. Dirección de Parques Nacionales y Vida Silvestre—Subsecreteria de Estado de Recursos Naturales y Medio Ambiente—Ministério de Agricultura y Ganadería, Asunción. Pp. 112–113.

Magalhães, C. and M. Türkay. 1996a. Taxonomy of the Neotropical freshwater crab family Trichodactylidae I. The generic system with description of some new genera Crustacea: Decapoda: Brachyura). Senckenbergiana biologica, 75:63–95.

Magalhães, C. and M. Türkay. 1996b. Taxonomy of the Neotropical freshwater crab family Trichodactylidae II. The genera *Forsteria, Melocarcinus, Sylviocarcinus,* and *Zilchiopsis* (Crustacea: Decapoda: Brachyura). Senckenbergiana biologica, 75:97–130.

Magalhães, C. and M. Türkay. 1996c. Taxonomy of the Neotropical freshwater crab family Trichodactylidae III. The genera *Fredilocarcinus* and *Goyazana* (Crustacea: Decapoda: Brachyura). Senckenbergiana biologica, 75:131–142.

Magalhães, C. and I. Walker. 1988. Larval development and ecological distribution of Central Amazonian palaemonid shrimps (Decapoda: Caridea). Crustaceana, 55:279–292.

Manning, R.B. and H.H. Hobbs, Jr. 1977. Decapoda. *In* Hurlbert, S.H. (ed.). Biota Acuática de Sudamérica Austral. San Diego: San Diego State University. Pp. 157–162.

Moreira, C. 1912. Crustacés du Brésil. Mémoires de la Société Zoologique de France, 25:145–154.

Nobili, G. 1896. Viaggio del Dott. A. Borelli nella Republica Argentina e nel Paraguay. Bollettino dei Musei di Zoologia ed Anatomia Comparata della Università di Torino, 11(222):1–4.

Omori, M. 1975. The systematics, biogeography, and fishery of epipelagic shrimps of the genus *Acetes* (Crustacea, Decapoda, Sergestidae). Bulletin Ocean Research Institute, 7:1–91. Tokyo: University of Tokyo.

Rathbun, M.J. 1906. Les crabes d'eau douce (Potamonidae). Nouv. Archs Mus. Hist. Nat., Paris 8:33–122.

Rodríguez, G. 1992. The freshwater crabs of America: family Trichodactylidae and supplement to the family Pseudothelphusidae. Collection Faune Tropicale, 31, Editions ORSTOM, Paris. 189pp.

Stimpson, W.W. 1861. Notes on certain decapod Crustacea. Proceedings of the Academy of Natural Sciences of Philadelphia, 13:372–375.

CHAPTER 6

Diversity and Abundance of Fishes in the Upper and Lower Río Paraguay Basin and the Río Apa Sub-basin, Paraguay

Mônica Toledo-Piza, Barry Chernoff, Darío Mandelburger, Mirta Medina, Jaime Sarmiento, and Philip W. Willink

Abstract

Two teams of ichthyologists sampled fishes at a combined total of 111 localities (Map 2). During the AquaRAP expedition, 136 fish species were collected in the Río Paraguay, 85 in the Río Apa, and 35 in the Riacho La Paz, for a grand total of 173 species.

Characiformes made up 54% of the fish fauna in the Río Paraguay. They belong to a variety of trophic categories, including herbivores, insectivores, mud-feeders, piscivores, and omnivores. Siluriformes (catfishes) and Perciformes (primarily cichlids) were the next most abundant groups. Piranhas (*Serrasalmus* spp.) were common and many fish exhibited piranha bites. Some fishes (e.g., pacus, anostomids) feed on fruits and seeds, and depend on floods to reach a plentiful supply of food. Flooded forests are often used as nursery areas by many species of fishes, including commercially important species. Many taxa (e.g., *Lepidosiren paradoxa*, *Synbranchus marmoratus*, Callichthyidae) possess adaptations for breathing air, which is important for survival through the dry season. Only the eggs of other species, known as annual fishes (e.g., *Rivulus* sp.), survive through the dry season.

The diversity of fishes was lower in the Río Apa than in the Río Paraguay, probably because of the lower habitat heterogeneity. Characiformes, Siluriformes, and Perciformes were still the dominant taxa. Many species collected in the Río Apa were not collected in the Río Paraguay, highlighting the distinctiveness of the two basins. Several species of Siluriformes were found amidst the cover of logjams. Schools of *Salminus, Brycon*, and *Prochilodus* were observed migrating upstream.

Introduction

Based on an intensive exploration of the ichthyofauna during the previous and present centuries, a series of publications on Paraguay fishes were produced (Perugia, 1891; Boulenger, 1895, 1897, 1900; Eigenmann and Kennedy, 1903; Eigenmann et al., 1907). Limited studies were conducted until the late 1980's, and it is estimated that approximately 600 fish species live in the La Plata drainage (Bonetto, 1986). The country of Paraguay is believed to contain 450 species (Lowe-McConnell, 1975:47), with the greatest diversity (300 species) in the north and decreasing to 200 species at the southern end (Bonetto, 1986). This high biodiversity in Paraguay as compared to neighboring countries (excluding Brasil) (Lowe-McConnell, 1975:47) can be partially attributed to two factors: (i) northern Paraguay is a continuation of the species rich Pantanal, and (ii) an apparent distinction between the Río Paraguay fauna and Río Paraná fauna, possibly due to the independent histories of the two rivers that may have later been combined into one drainage by a "capture event" (Lowe-McConnell, 1975:48).

Since approximately 1990, the Museo Nacional de Historia Natural del Paraguay (MNHN-P), the Geneva Museum, and the Swedish Museum of Natural History, Stockholm (NRM), have been maintaining permanent programs in Paraguay. Although these programs are directed towards the knowledge of general fauna, they have produced a significant volume of information about ichthyology. Collection efforts made by these programs (and earlier collectors) usually concentrated on the area of the Río Paraguay and its tributaries between the mouths of Ríos Apa and Paraná. Samples originating from the northern Río Paraguay are much more scattered and come from few localities along this portion of the river. Consequently the collections made in the northern Río Paraguay by the AquaRAP expedition represent an important contribution to the knowledge of

Paraguayan ichthyology, by providing information from previously poorly collected areas and thus complementing information from other collections. It is imperative to have this information in order to make qualified conservation and management decisions.

Methods

During the AquaRAP expedition, 15 prospective stations were established (10 in the Río Paraguay (including one in the Río Negro), two in the Río Apa, three in the Riacho La Paz and small streams in the region close to Concepción), from which a total of 111 fish localities were sampled (Appendix 12, Map 2). The period of stay at each station was generally 24 hours, except in the Río Apa zone where the period of stay was three days.

Two working teams composed of two ichthyologists, one assistant, and a motorist were established for the fish sampling operations, and each team had a motorboat. Sampling in the Río Paraguay proper was very limited due to unusually high water levels. This resulted in a decreased availability of both beaches and appropriate areas for trawling.

In the few shallower areas (beaches, riparian forests, or some palm stands), collecting was accomplished with seines of different sizes (approximately 5 meters long by 1.8 meters high, with small mesh). Seining was executed in areas free of vegetation, but more frequently in areas with floating meadows, palm stands, or flooded forests. Occasionally, due to the high water level, the seines were stretched between two boats and dragged along the surface, enclosing a portion of the floating vegetation, which was then hoisted to one of the boats for detailed inspection. In other cases, the nets were dragged to the beach with the aid of one of the boats.

Gill nets 42 meters long by 2 meters high, with five mesh sizes between 10 to 60 millimeters, were used. The nets were usually set at dawn and collected between 6–8 am the following day. Nets were checked once at night and then in the following morning. Gill nets were set in areas of flooded forests, flooded palm stands, and coves that usually bordered floating meadows (*Eichhornia* sp. and Cyperaceae).

In the main river channel (depth > 5 meters) we used a trawl with an anterior opening of 1 by 3 meters, 20 millimeter mesh, in which the posterior end had a fine mesh collection sleeve. Runs were done with a motorboat over clean bottoms (free from debris) for 5 minute periods. Hook and line (fishing rod) was also utilized, mainly near the edges of the river and next to floating meadows. In some shallower areas with vegetation, hand nets (entomological nets) were used.

Fishes were preserved in 10% formalin solution. All specimens captured at the same place and time were maintained separately from all other collected specimens. All material was wrapped and shipped to Chicago for sorting, identifica-

tion, and enumeration in the Division of Fishes, Department of Zoology, Field Museum (FMNH). Fifty percent of the specimens are housed in the MNHN-P, San Lorenzo, Paraguay; the remaining specimens were shared among the participating institutions: FMNH; Museu Zoologia do Universidade de São Paulo, São Paulo, Brasil; and Museo Nacional de Historia Natural, La Paz, Bolivia.

The identifications were made in a careful but relatively rapid fashion. General works such as Eigenmann and Myers (1929) or Géry (1977) were used but preference was always given to systematic revisions (e.g., Vari, 1992; Mago-Leccia, 1994) and recent species descriptions (Mahnert and Géry, 1988), if available. In many cases specimens were compared to types or historic material referenced in the literature and housed at the FMNH. However, identification to the level of species or even genus was not always possible. To do so would represent a less than scholarly approach to the taxonomy. Instead we rely upon morphospecies—the number of distinguishable entities present in our samples. This bears the assumption that such discernable entities or morphospecies are putative taxonomic entities (i.e., species). We were careful to check for sexual and ontogenetic differences. All specimens were examined critically and identified to their lowest taxonomic level (Appendix 13).

Río Paraguay

The area studied included a portion of the Río Paraguay, between Bahía Negra on the Bolivian border and Concepción. The northern extent was the Río Negro, near its outlet into the Río Paraguay. The southern extent was the Río Aquidabán and Río Napegue. In the region of Riacho Mosquito (near the Río Apa outlet), some sampling was done in Chaco streams to the west of the Río Paraguay. Collecting in the Río Paraguay was difficult due to unusually high water levels, resulting in a decreased availability of both beaches and appropriate areas for trawling.

During the AquaRAP expedition, 136 different fish species were collected in the Río Paraguay (Appendix 13). This corresponds to 79% of the species collected during the entire expedition. Of the 136, 74 (54%) are Characiformes, 40 (29%) are Siluriformes, and 11 (8%) are Perciformes. Similar to the results for the entire expedition, most of the species collected in the Río Paraguay belonged to the family Characidae (56), followed by Loricariidae (17), Pimelodidae (12), and Cichlidae (11).

Characiformes included representatives from a wide range of trophic categories: such as herbivores, insectivores, mud-feeders, piscivores, and omnivores. The most common species in our collection were the characid species *Holoshesthes pequira*, *Hyphessobrycon eques*, *Moenkhausia dichroura*, *M. sanctaefilomenae*, *Aphyocharax anisitsi*, *A. dentatus*, and *Astyanax paraguayensis*. They are all small-bodied species and were collected in shallow water habitats such as beaches, backwaters, and flooded forests.

Piscivores included the toothed predators *Acestrorhynchus pantaneiro* and *Rhaphiodon vulpinus*. The latter species exceeded 70 centimeters in length and was collected in the main channel of the Río Paraguay. *Acestrorhynchus pantaneiro* was collected in areas of standing water in flooded forests and lagoons. Piranhas (*Serrasalmus*) constituted the largest number of specimens collected in the main channel of the Río Paraguay. Many of the fishes collected with gill nets had their bodies damaged to varying extents by piranha bites, indicating a high density of piranhas in the area. Fishes of the family Erythrinidae, represented by the predators *Hoplias malabaricus* and *Erythrinus erythrinus*, were collected in different habitats such as beaches, lagoons, and flooded forests. The large food and sport fish, *Salminus maxillosus*, is a piscivore that was restricted in our collections to the main channel.

Mud-feeders are represented by the sábalo (carimbatá) (*Prochilodus lineatus*), a dominant species in the Río Paraguay and important source of human food (Mago-Leccia, 1972 in Lowe McConnell, 1975:113). In some areas they constitute between 50 and 60% of the total fish biomass (Bonetto, 1986). Other mud-feeders include members of the family Curimatidae, represented in this study by the genera *Curimatella*, *Potamorhina*, *Psectrogaster*, and *Steindachnerina*. Curimatids were collected in the majority of the habitats sampled. The mud-feeders play an important role in speeding the nutrient recycling and increasing the productivity in the habitats where they occur, acting in the initial phases of the mineralization of organic matter and providing bacteria with a substrate easier to decompose (Gneri and Angelescu, 1951 in Agostinho et al., 1997).

Insectivores included the relatively small species of the genera *Astyanax*, *Moenkhausia*, and *Roeboides*. They play an important role as forage for the piscivores (Agostinho et al., 1997). *Roeboides paranensis* is interesting for having many mammiliform teeth entirely external to the mouth and pointed forward, a specialization for scale-eating. Arthropods are ingested in order to complement their diet (Sazima, 1983).

The pacus, represented by species of the genera *Myleus*, *Metynnis*, and *Mylossoma*, are herbivores feeding on fruits and seeds. Also present in the study area is the pacu, *Colossoma mitrei* (collected by fisherman, not present in our collection), which can reach large sizes and are considered excellent eating. Species of *Leporinus* and *Schizodon* (family Anostomidae) also feed largely on fruits, seeds, leaves, and algae. *Leporinus friderici* also feeds on fishes, although to a minor extent, and appears to be omnivorous (Goulding, 1980:172; Agostinho et al., 1997).

The Siluriformes comprise 30% of the collected species being represented by the families Ageneiosidae, Aspredinidae, Auchenipteridae, Callichthyidae, Doradidae, Loricarii-

dae, and Pimelodidae. Some species are of commercial value, having been seen in the local market at Concepción, and include species of the family Pimelodidae such as *Paulicea lutkeni*, *Pseudoplatystoma coruscans*, *P. fasciatum*, and Doradidae, such as *Oxydoras kneri* and *Pterodoras granulosus*.

Fishes of the family Loricariidae are bottom dwelling catfishes living on the substrate, which may take the form of submerged tree trunks, rocks, pebbles, sand, or even muddy bottoms of still waters. They are flattened and almost entirely covered by bony scutes and have ventral, suctorial mouths that, along with a set of rasping teeth, provide the means by which they are able to feed on algae, detritus, small crustaceans, and insect larvae. Nine loricariid genera were collected in Río Paraguay. *Otocinclus* was the most common loricariid, represented by three species in our collection. They are small herbivorous catfishes, most frequently living in close association with aquatic macrophytes and terrestrial marginal grasses extending into the water column. They are typically found in locations characterized by clear, well oxygenated water of moderate flow (Schaeffer, 1997). They were usually present in relatively large number of individuals, along the Río Paraguay, in habitats such as beaches, backwaters, flooded forests, and under floating vegetation. *Liposarcus anisitsi* (zapato) was the biggest loricariid collected with a large number of individuals collected in the main channel of the Río Negro, near its outlet into Río Paraguay. These fish can reach significant sizes and play an important role in subsistence fishing. *Rineloricaria*, *Hypoptopoma*, and *Loricariichthys* were also relatively frequent being found in shallow water habitats such as beaches, backwaters, flooded forests, and also under floating vegetation.

Fishes of the family Callichthyidae are usually small in size and have the ability to supplement their oxygen intake by gulping atmospheric air and swallowing it into their guts where the oxygen is extracted. This process allows callichthyids to survive in waters with little dissolved oxygen and increase their oxygen consumption during periods of stress, even allowing them to move out of the water when searching for favorable environments. The diet of most species consists of worms and insect larvae rooted out of soft sandy or muddy material with the use of their barbels. They are represented in our collection by *Corydoras hastatus* and *Megalechis thoracata*.

Ostariophysans are also represented by the Gymnotiformes, characterized by the presence of electric organs and elongate bodies with a long anal fin. They are mostly nocturnal fishes, some being crepuscular such as *Eigenmannia* (the most abundant gymnotiform in our collections). They feed mostly on aquatic insects (particularly Chironomidae) and smaller crustacea (Mago-Leccia, 1994:14).

Cyprinodontiformes are usually small, and some species, characterized by the development of their fins and colors, are quite important to aquarists and commercial trade (e.g., *Cynolebias*, *Pterolebias*). These species are usually associated with marshes habitats, of which we sampled only one example during the expedition (Riacho Mosquito, a Chaco stream that feeds into the Río Paraguay south of the mouth of the Río Apa), and captured one specimen of *Rivulus* sp. Within this group we find the aforementioned annual fishes, which with the first rain will develop quickly and reproduce before the next season.

Synbranchidae is represented in the collection by *Synbranchus marmoratus*. This eel-like fish is most common in lentic situations and is capable of withstanding low oxygen levels by "breathing" air. Under extreme conditions, they are capable of moving over land to search for a more hospitable locality.

Perciformes are represented by the Cichlidae. Cichlids, popular among aquarists, were usually collected in lentic systems. The largest species is the apaiari (serepapa) (*Astronotus crassipinnis*). Species of *Apistogramma* (*A. borellii*, *A. commbrae*, and *A. trifasciata*) were the most abundant, followed by *Crenicichla lepidota* and *Cichlasoma dimerus*. The diet of *Apistogramma* and *Cichlasoma* includes a wide spectrum of items such as, plants, fruits, and insects, with the diet of *Crenicichla* tending to be more carnivorous (Knöppel, 1970).

Lepidosiren paradoxa is the only species of lungfish in the neotropics. They were usually associated with the vegetation in floating meadows, and often found over muddy bottoms. Lungfish are dependent on atmospheric oxygen, and are capable of building burrows in which to aestivate when water levels drop. These adaptations allow them to survive the periodic flooding and drying of the wetlands.

Cartilaginous fishes were represented by the stingrays (*Potamotrygon* spp.). The spine in their tails can cause painful wounds that take a long time to heal, and they have been reported to reach 1.5 m in disc diameter (Bonetto, 1986).

Few species are represented in our collection by only one individual, including the pimelodids *Hemisorubim platyrhynchos*, *Sorubim lima*, *Pinirampus pirinampu*, and the Ageneiosidae species *Ageneiosus* cf. *brevifilis*. These species, together with *Salminus maxillosus*, *Rhaphiodon vulpinus*, *Plagioscion* sp., and *Pellona* sp., were only collected in the main channel of the Río Paraguay. Their apparent limited abundance was most likely due to the use of collection methods inappropriate for these particular species. Juveniles of a few other species collected in the main channel, such as *Serrasalmus marginatus*, *Hypostomus* sp., and *Mylossoma* sp., were collected in shallow water habitats—flooded forests, for example. Species at different stages of development were collected, such as *Lepidosiren paradoxa*, *Acestrorhynchus pantaneiro*, *Synbranchus marmoratus*, and *Hoplias malabaricus*. In one lagoon at Estancia Cerrito (21° 27' S; 57° 55' W), fish larvae were associated with the roots of the aquatic plant *Eichhornia*. Nakatani et al. (1997) studied the ecology of fish larvae in the Río Paraná basin, and pointed out the importance of flooded areas for the initial developmental stages and growth of many fish species.

Río Apa

In the lower part of the Río Apa, some collections were made between the outlet into the Río Paraguay and the rapids. The middle part was sampled between the outlet of the Riacho Blandengue and the rapids in the San Carlos area. To a lesser degree, samples were collected in some tributaries like Riacho Blandengue and Riacho Toro Paso.

During the expedition, 85 different fish species were collected in the Río Apa (Appendix 13). Of these, 49 are Characiformes (58%), 28 are Siluriformes (33%), and 4 are Perciformes (5%). Most of the species collected belong to the families Characidae (37), Loricariidae (15), Pimelodidae (7), and Cichlidae (4).

Several of the species collected in the Apa had not been collected in the Paraguay, despite the intensive work previously done in the Paraguay. They are probably species restricted to smaller systems. A similar result can be seen in the University of Michigan Museum of Zoology (UMMZ) and NRM collections. Among the species collected by AquaRAP only in the Río Apa, some are represented by very few specimens (5). These include the characiforms *Leporinus friderici acutidens*, *L. striatus*, *Clupeacharax anchoveoides*, *Jupiaba acanthogaster*, and *Thorachocharax stelatus*; and the siluriforms *Bunocephalus doriae*, *Tatia aulopigia*, *Callichthys callichthys*, *Corydoras aeneus*, *Corydoras ellisae*, *Pseudcetopsis gobioides*, *Pimelodella laticeps*, and *Imparfinis minutus*. The latter species was collected at the region of the rapids above San Carlos, on rocks in the middle of the river subject to strong currents.

Other species were collected in relatively higher numbers, such as *Piabarchus analis*. The genus *Piabarchus* was poorly known until recently, being rare in collections. Based on material collected by the joint programs between the MNHN-P and the Geneva Museum, Mahnert and Géry (1988) redescribed *P. analis* and described a new species, *P. torrenticula*. The genus is restricted to the Río Paraguay basin.

Xenurobrycon macropus is the only representative of the Glandulocaudine collected in the present study, and it is one of the most interesting examples of miniaturization. Maximum known body length is around 20 millimeters and they reach sexual maturity at 12 millimeters (Weitzman and Vari, 1988). Glandulocaudines are also interesting for having internal fertilization. The males of all glandulocaudine species have caudal glands and certain scales on the tail modified as different kinds of pumping mechanisms. The caudal glands are assumed to produce pheromones that are dispersed through the pumping mechanisms during courtship (Weitzman and Fink, 1985). In one of the localities over 500 specimens of *Xenurobrycon macropus* were collected at once.

Other common fishes in our collection were representatives of the genera *Astyanax*, *Moenkhausia*, *Aphyocharax*, and *Holoshesthes*.

Logjams provide cover for several species of Siluriformes (Auchenipteridae (*Tatia*), Doradidae, and Trichomycteridae) and others. Some species hide during the day in submerged branch and tree cavities, then come out at night to feed. On some occasions, *Rineloricaria* and *Ancistrus* with eggs or juveniles were collected among these logjams. Noteworthy in the case of the *Rineloricaria* is that a female and a male were found in the same cavity together (with all their secondary sexual characteristics well developed) protecting a litter of offspring.

Migratory movements

Of particular interest were the migratory movements we observed in the Río Apa close to the outlet of Riacho Blandengue. At this location, the river was approximately 100 m wide, not very deep (we were able to stand in at least some places in the middle of the river), the current was relatively strong, and there were many partially submerged logs and tree trunks. Around 12pm, schools of *Salminus*, *Brycon*, and *Prochilodus* were seen swimming upriver in apparent migratory movements.

Migrations are commonly observed in the Ríos Paraguay and Paraná, particularly of *Salminus maxillosus* and *Prochilodus lineatus*. These migrations can be 1000 kilometers long, and typically begin at the start of the rainy season (Bonetto, 1986). Reproduction is believed to be the impetus for most migrations in most species. Other species, such as *Pseudoplatysoma coruscans* and *P. fasciatum*, undergo multiple migrations. Some of these migrations are apparently related to reproduction (occurring during flood periods), while other migrations appear to be related to feeding (Bonetto, 1986). There is still much to be learned about migratory phenomena.

Riacho La Paz

Four samples were made in the Riacho La Paz near the outlet into the Río Paraguay, and one in the upper part of the stream on the route to San Carlos del Apa. A total of 35 fish species were identified for this drainage, 25 of which were Characiformes, 7 Siluriformes, 1 Cichlidae, 1 Belonidae, and 1 Potamotrygonidae (Appendix 13). One species of *Characidium* and two species of *Astyanax* (*A. fasciatus* and *A. lineatus*) occur in the upper portion of the river. The latter two species were also collected in the upper portions of Riacho Blandengue. The Riacho Blandengue possesses features similar to the Riacho La Paz, indicating that these species may be restricted to small stream habitats and areas close to headwaters.

Discussion

The 173 fish species collected during the AquaRAP-Paraguay expedition represent approximately 38% of the fish fauna currently estimated for the country of Paraguay (Lowe-McConnel, 1975:47). If we take into consideration the short duration of the expedition (approximately two weeks), the limited amount of time spent at each collecting station, and the collecting methods used, both of which did not allow us to sample every available habitat or conduct exhaustive surveys, then we conclude that the stretch of the river basin sampled (from Bahía Negra to Concepción) possesses a good representation of the ichthyofauna of the country. The species include representatives of all major freshwater fish families that occur in the neotropics, belong to a wide array of trophic categories, and occupy different habitats.

Although we collected 173 taxa, the total number of species in the region would rise with continued sampling. This conclusion is based on the species accumulation curve for the expedition, which never reached an asymptote (Fig. 6.1). There appears to be the beginning of a leveling off after nine days, but then the slope of the curve increases. This rate increase is concordant with the expedition's move from the Río Paraguay to the Río Apa. Habitats and species in the Apa were distinct from the Paraguay, adding to the overall diversity of the region. New species were also collected when the expedition returned to the Río Paraguay (Day 13+) (Fig. 6.1). Even after two weeks of sampling, species were being added at an average rate of eight species per day.

Among the collected species, approximately 22 are restricted to the Río Paraguay basin, with the remaining majority also occurring in other river basins such as the Ríos Paraná and Guaporé (Pearson, 1937) and Amazon basins (Appendix 14). Some of the widespread species belong to

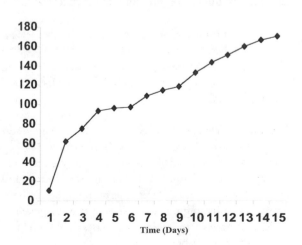

Figure 6.1. Species accumulation curve for fishes collected during the Río Paraguay AquaRAP expedition, 4–18, September, 1997.

poorly studied genera and many of them probably constitute species complexes. Revisionary studies on these groups may reveal the existence of more species endemic to the Río Paraguay basin. In addition, some of the taxa identified as sp. 1, 2, etc. (Appendix 13) may represent new, yet undescribed, species.

Although we have not performed an exhaustive survey of the fish fauna to allow a detailed comparison among different habitats and portions of the sampled area, some informative preliminary observations were obtained. Examples are: differences in the species collected in the Ríos Apa, Paraguay, and headwater streams (Riacho La Paz), and differences both taxonomic and in the life stages of the individuals collected in the various habitats (e.g., flooded forest, main channel, lagoons, rapids, log jams, and headwater streams). The latter indicates that specific habitats are critical for the survival of specific stages during a species life cycle reinforcing the importance of the conservation of areas that includes a wide representation of habitats in order to insure the survival of the species.

The migratory movements observed also show the high dependency of the fish fauna on the hydrological cycle of the basin as a whole. Currently there is no available information about migration patterns of the fish populations in the area we studied. Although detailed studies are necessary our observation of migration movements of a few commercially important species in the Río Apa indicates that this river probably plays an important role in the life cycle of those fish populations. As emphasized by Goulding (1980), it is important to have a broad view of the fish fauna as opposed to seeing the fishes and their environment within the restricted and static framework of a particular area or time period.

Recommended Conservation and Research Activities

• Assess the reproductive and growth rates of commercial species of fishes. Determine levels of sustainable yield and adjust legislation accordingly. Monitor catches of fishes. International cooperation is required because fishes cross country borders.

• Aquarium fishes (e.g., cichlids, tetras [characids], armored catfish [loricariids]) were collected in the study area. Unrestricted large-scale harvesting of these species would destroy their populations, as has occurred in some other areas of South America. After determining population sizes and conducting impact studies, it may be possible for local fishermen to implement a sustainable yield fishery for aquarium fishes.

• Migrations of *Salminus*, *Brycon*, and *Prochilodus* were observed in the Río Apa; hence this river appears to play

an important role in the life history of these commercially important fishes. Studies need to be conducted to learn more about these migrations. Regulations that protect these migrations need to be created and enforced.

• Habitats and species in the Río Apa are distinct from the Río Paraguay. Portions of both areas need to be managed in order to maximize the conservation of the entire region's biodiversity.

• Within a species, individuals at different life stages were collected in the various habitats (e.g., flooded forest, main channel, lagoons, rapids, log jams, and headwater streams). This indicates that specific habitats are critical for the survival of specific stages during a species life cycle reinforcing the importance of the conservation of areas that includes a wide representation of habitats in order to ensure the survival of the species.

• Prevent large- and small-scale disruptions of the seasonal flooding. Disruptions of flooding would result in the loss of wetlands (i.e., nursery areas for fishes, etc.) and devastate biodiversity.

Literature Cited

Agostinho, A. A., N. S. Hahn, L. C. Gomes, and L. M. Bini. 1997. Estrutura trafica. *In* Vazzoler, A. E. A. M., A. A. Agostinho, and N. S. Hahn (eds.). A planície de inundagco do alto Rio Paraná: aspectos físicos, biológicos e socioeconômicos. Editora da Universidade Estadual de Maringa. Pp. 229–248.

Bonetto, A.A. 1986. Fish of the Paraná system. *In* Davies, B.R. and K.F. Walker (eds.). The ecology of river systems. Dr. W. Junk Publishers, Dordrecht, Netherlands. Pp. 573–588.

Boulenger, G.A. 1895. Description of two new South American characinoid fishes. Annals and Magazines of Natural History, 6:9–12.

Boulenger, G.A. 1897. Viaggio del Dr. Alfredo Borelli nel Chaco boliviano e nella Republica Argentina. III. Poissons. Bollettino dri Musei di Zoologia ed Anatomia Comparata della Università di Torino, 12:1–4.

Boulenger, G.A. 1900. Viaggio del Dott. Alfredo Borelli nel Mato Grosso e nel Paraguay. III. Liste des poissons recueillis a Urucum et a Carandasinho, pres de Corumba. Bollettino dri Musei di Zoologia ed Anatomia Comparata della Universit. di Torino, 15:1–4.

Eigenmann, C.H. and C.H. Kennedy. 1903. On a collection of fishes from Paraguay, with synopsis of the American genera of cichlids. Proceedings of the Academy of Natural Sciences of Philadelphia, 55:497–537.

Eigenmann, C.H., W.L. McAtee, and D.P. Ward. 1907. On further collections of fishes from Paraguay. Annals of the Carnegie Museum, 4:110–157.

Eigenmann, C.H. and G.S. Myers. 1929. The American Characidae. Memoirs of the Museum of Comparative Zoology, 43:429–574.

Géry, J. 1977. Characoids of the world. Neptune City, New Jersey: T.F.H. Publications. 672pp.

Goulding, M. 1980. The fishes and the forest. Berkeley: University of California Press. 280pp.

Knöppel, H.-A. 1970. Food of central amazonian fishes: contribution to the nutrient-ecology of amazonian rain-forest-streams. Amazoniana, 2:257–352.

Lowe-McConnell, R. H. 1975. Fish communities in tropical freshwaters: their distribution, ecology and evolution. New York: Longman. 337pp.

Mago-Leccia, F. 1994. Electric fishes of the continental waters of America. Fundacion para el Desarrollo de las Ciencias Fisicas, Matematicas y Naturales (FUDECI) Vol. XXIX. Caracas, Venezuela. 206pp.

Mahnert, V. and J. Géry. 1988. Les genres *Piabarchus* Myers et *Creagrutus* Günther du Paraguay, avec la description de deux nouvelles espèces (Pisces, Ostariophysi, Characidae). Revue Francaise d'Aquariologie, 15:1–8.

Nakatani, K., G. Baumgartner, and M. Cavichioli. 1997. Ecologia de ovos e larvas de peixes. *In* Vazzoler, A. E. A. M., A. A. Agostinho, and N. S. Hahn (eds.). A planície de inundagco do alto Rio Paraná: aspectos físicos, biológicos e socioeconômicos. Editora da Universidade Estadual de Maringa. Pp. 281–306.

Pearson, N.E. 1937. The fishes of the Beni-Mamoré and Paraguay basins, and a discussion of the origin of the Paraguayan fauna. Proceedings of the California Academy of Science, 23:99–114.

Perugia, A. 1891. Appunti supra alcuni peci Sud-americani conservati nel Museo Civico di Storia Naturale di Genova. Annali Museo Civico di Storia Naturale di Genova, 10:605–657.

Sazima, I. 1983. Scale-eating in characoids and other fishes. Environmental Biology of Fishes, 9:87–101.

Schaeffer, S. 1997. The Neotropical cascudinhos: systematics and biogeography of the *Otocinclus* catfishes (Siluriformes: Loricariidae). Proceedings of the Academy of Natural Sciences of Philadelphia, 148:1–120.

Vari, R.P. 1992. Systematics of the Neotropical characiform genus *Curimatella* Eigenmann and Eigenmann (Pisces: Ostariophysi) with summary comments on the Curimatidae. Smithsonian Contributions to Zoology, 533:1–48.

Weitzman, S. H. and S. V. Fink. 1985. Xenurobryconin phylogeny and putative pheromone pumps in glandulocaudine fishes (Teleostei: Characidae). Smithsonian Contributions to Zoology, 421:1–120.

Weitzman, S. H. and R. P. Vari. 1988. Miniaturization in South American freshwater fishes: an overview and discussion. Proceedings of the Biological Society of Washington, 101:444–465.

CHAPTER 7

Testing Hypotheses of Geographic and Habitat Partitioning of Fishes in the Río Paraguay, Paraguay

Barry Chernoff, Philip W. Willink,
Mônica Toledo-Piza, Jaime Sarmiento,
Mirta Medina, and Darío Mandelberger

Abstract

The 173 species of freshwater fishes collected during the AquaRAP expedition to the Río Paraguay were analyzed to determine if any distributional patterns existed within the region collected. The region was divided into five subregions and also into 10 macro-habitats. Two null hypotheses were tested and rejected: (i) that fishes are randomly distributed with respect to sub-region; and (ii) that fishes are randomly distributed with respect to macro-habitat. The results show that a strong subregional effect was evident such that two distributional elements were present: an association between the Río Paraguay and Río Negro subregions, and an association between the Río Apa and Riacho La Paz subregions. The analysis of distributions with respect to macro-habitats also reveals two components. The first relates the beach and main channel faunas to the macro-habitats that are inundated during seasonal flooding, such as flooded forests and lagoons. The assemblage associated with the flooding cycle accounts for more than 75% of the fishes collected. The second component comprises habitats found within the Río Apa and the Riacho La Paz (e.g., clear water, rapids, etc.). This Río Apa-Riacho La Paz component has a relatively sharp boundary with respect to the Río Paraguay with more than a 50% turnover in fauna. These results show that core conservation areas must be set up within situations in which the seasonal cycle of flooding is unimpeded and the area of inundation is relatively unmodified. Also, the Río Apa-Riacho La Paz represents a highly threatened area because of high rates of land conversion in the region, aridity, and the fact that much of the fauna cannot be recolonized from nearby water sources.

Introduction

The fishes of the Río Paraguay comprise a rich assemblage with more than 250 species documented from the entire river basin (Bonetto, 1986; Lowe-McConnell, 1987). Toledo-Piza et al. (2001) collected 173 species in the Río Paraguay between Concepción and the Río Negro. Based on observations made during the expedition, we were able to divide the region of the AquaRAP survey into five subregions: Río Negro, Upper Río Paraguay, Lower Río Paraguay, Río Apa and Riacho La Paz (Map 2). Furthermore, we were able to recognize 10 macro-habitats (e.g., beaches, flooded forests, rapids, etc.) within which the fishes were captured. There are many trenchant threats to the aquatic ecosystems of the Río Paraguay (e.g., Hidrovia, land conversion, etc.) and this paper will explore what the consequences of the threats might be given the distributions of fishes with respect to the subregions and macro-habitats.

A pattern of heterogeneous distribution by the fishes within the Río Paraguay with respect to subregions and macro-habitats would have important ramifications for conservation recommendations. For example, if the fauna were homogeneously distributed then a core conservation area could be established that might effectively protect the vast majority of the species. However, as the distribution of the species either among subregions or among macro-habitats becomes increasingly distinct and patchy, then a single core area, apart from the entire region, may not provide the desired level of protection. Chernoff et al. (1999) demonstrated how we can use information on the relative heterogeneity of distributions among sub-regions or among macro-habitats to predict possible faunal changes in response to specific environmental threats and that such analyses be carried out within the framework of a rapid assessment program.

This paper will test two null hypotheses that are critical to the conservation of freshwater fishes of the Río Paraguay as follows: that the fishes are randomly distributed among (i) five subregions; and (ii) 10 macro-habitats. To test these hypotheses we add to the relatively new approaches of Chernoff et al. (1999) as well as use methods developed by Atmar and Patterson (1993). Because the two null hypotheses are rejected, we then estimate what changes in the ichthyofauna might take place given particular environmental threats.

Methods

The basic data used to test these hypotheses were lists that correspond to that shown for the five subregions in Appendix 13. Although we counted all of the individuals captured at each field sampling station, we make no calculations of abundance or in any way use the number of times that a species was collected in a subregion, in a macro-habitat or in a type of water. We scored the presence of a species as "1" and its absence as "0" in matrices with species on the rows and subregion or macro-habitat on the columns.

Given point source data obtained in rapid assessments only the presence of a species should be used as information. As discussed in Chernoff et al. (1999), the absence of a species is ambiguous. Absence of a species signifies either of the following: (i) it was not collected in a sub-basin or macro-habitat; or (ii) the species does not live in that sub-basin or habitat; given the nature of rapid assessments we cannot distinguish between the alternatives. In the methods below, only the presence of a species constitutes data.

The results of the methods that follow are potentially influenced by relative "effort". Because we are not using abundance information, "effort" refers to the number of field stations (collections) that pertain to each categorical variable of subregion or macro-habitat. If, for example, we found more species present in subregion or macro-habitat "A" than in subregion or macro-habitat "B", this result could be due to the fact that more collections were made in "A" than in "B". This is especially important because Toledo-Piza et al. (2001: Fig. 6.1) demonstrated that the species accumulation curve was not asymptotic—the number of species continued to rise with the number of collections made. To determine if there is an effect of sampling effort relative to subregion or to macro-habitat, the number of species found in each subregion or macro-habitat was regressed onto the number of collection stations. Because the independent variable, number of collection stations, was determined without error, least squares regressions were used (Sokal and Rohlf, 1995).

Chernoff et al. (1999) selected Simpson's Index of Similarity, Ss, as the most consistent with data collected during rapid inventories or with point source data. Simpson's

Index uses the following table format to calculate the similarity between two lists or samples of species:

		Sample 1	
		1	0
Sample 2	1	a	b
	0	c	d

where, a is the number of positive matches or species present in both samples, b is the number of species present in sample 2 and absent from sample 1, c is the converse of b, and d is the number of negative matches or species absent from both localities. Simpson's Index of Similarity, $S_s = a/(a+b)$, where $b < c$, or $S_s = a/n_s$ where n_s is the number of species present in the smaller of two lists. The denominator of the index eliminates interpretation of the negatives – absent species. The 0's in the matrices are really coding artifacts or place holders for missing data.

In order to interpret the observed similarity of two samples, both of which are drawn from a fixed larger universe (e.g., the set of all species captured during the AquaRAP expedition to the Río Paraguay), we undertake a four step procedure. In **step 1,** we calculate S_s by reducing via rarefaction the number of species in the larger sample to equal the number of species in the smaller sample, n_s. This rarefaction and calculation of S_s is iterated 200 times. From the 200 simulations a mean similarity, S'_s, is calculated and reported in tables of similarity (e.g., Table 7.5). This procedure is repeated to calculate an S'_s for each pair of samples in the analysis (e.g., Table 7.5).

Interpreting the significances of the mean similarities among the samples requires simulations across the range of number of species found in the samples of subregions and macro-habitats. In **step 2,** we simulate 200 random pairs of samples by bootstrapping with replacement from the set of all species captured during the expedition with the constraint that each random sample contains a fixed number of species

Table 7.5. Means of Simpson's Index of Similarity, S'_s, among ichthyofaunas found in five subregions of the Río Paraguay. Larger samples were rarefied 200 times to the size of the smaller samples. Index values are reported as percentages. ** indicates P<0.0001. Abbreviations: APA–Río Apa; LP–Lower Río Paraguay; RLP–Riacho La Paz; RN–Río Negro; UP–Upper Río Paraguay.

	UP	LP	APA	RLP
LP	69.6**			
APA	54.0	52.9		
RLP	58.8**	64.7**	85.3**	
RN	80.6**	72.2**	52.8**	29.4

for a given point in this range. For each random pair of the 200 we calculate their Simpson's Similarity. These 200 random similarities approximate a normal distribution (Fig. 7.1) from which we calculate a mean and standard deviation due to random causes; henceforth called mean random similarity, S^*_n, where n refers to the number of species present in the sample. For example, based upon the distribution shown in Fig. 7.1, S^*_{70} is 40.91 with an observed standard deviation of 4.63.

Random similarity distributions were generated at intervals of 10 species in order to estimate S^*_n and its standard deviation for samples containing between 20 and 120 species. This range of random list-sizes encompasses the actual number of species observed in subregions and in macro-habitats. In **step 3,** the means and standard deviations

are plotted against number of species present in a sample (Figs. 7.2–3). As the number of species present in a sample increases the observed similarity due to random effects also increases (Fig. 7.2) but the variance decreases (Fig. 7.3).

In **step 4,** we compare the observed mean similarity, S'_s, calculated from rarefaction (step 1) to the predicted value of S^*_n and its standard deviation from the regressions presented in Figures 7.2–3 (step 2). Using a 2-tailed parametric approach, the probability of obtaining the observed similarity at random is calculated from the number of standard deviations that the observed similarity was either above or below the mean of the bootstrap random distribution. This probability was obtained by interpolation of the values presented in Rohlf and Sokal (1995: Table A). The significance of the probability values was adjusted with the sequential Bonferroni technique (Rice, 1989) because each sample is involved in multiple comparisons. The sequential Bonferroni procedure is conservative, making it harder to reject a null hypothesis. We selected the P=0.01 level as our criterion for rejection of a null hypothesis. The value, 0.01, was divided by the number of off-diagonal comparisons present in the upper or lower triangle of the matrix of observed similarities. This new result is used as the criterion to evaluate the null hypothesis that $S'_s = S^*_n$. For example, in the lower triangle of Table 7.2 there are 10 similarities. In order to reject the null hypothesis, S'_s must be more than 3.1 standard deviations above or below S^*_n so that P<0.001.

If S's is found to be significantly different from S^*_n, then we reject the null hypothesis and conclude that the observed *similarity* is not due to random effects. However, if S'_s falls within the random effects or if S'_s is greater than the random mean, we fail to reject the null hypothesis concerning the *samples*—that the two *samples* are equal. In the former case we conclude that the two lists are drawn homogeneously from a larger distribution. In the latter case, we conclude

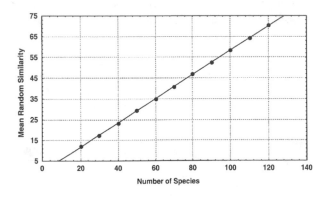

Figure 7.1. Distribution of Simpson's Similarity Index for 200 randomly simulated vectors of presence-absence data given the constraint that the universe size is 173 species and the constraint that 25 species are present in the sample.

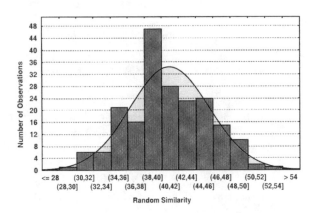

Figure 7.2. Mean random similarity, Simpson's Index, plotted against the number of species constrained to be present in the sample. Each mean was calculated from 200 random samples for populations of a given number of species.

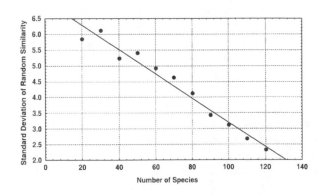

Figure 7.3. Standard deviation of random similarity, Simpson's Index, plotted against the number of species constrained to be present in the sample. Each standard deviation was calculated from 200 random samples of a given number of species.

Table 7.2. The number of species shared (lower triangle) and the number of species shared uniquely (upper triangle) among the subregions within the Río Paraguay. Abbreviations: APA–Río Apa; LP–Lower Río Paraguay; RLP–Riacho La Paz; RN–Río Negro; UP–Upper Río Paraguay; n–total number of species; u–number of uniques species; %u–percentage of unique species to number of species within a subregion.

	UP	LP	APA	RLP	RN
UP		19	5	0	1
LP	64		5	0	1
APA	47	46		6	0
RLP	20	22	29		0
RN	29	26	19	10	
n	105	93	85	35	41
u	27	18	24	4	6
%u	25.7	19.4	28.2	11.4	14.6

that the similarity is due to biological dependence or correlation, such as nested subsets (i.e., one population forms the source population for another). If S'_s is significantly less than S^*_n mean, we reject the null hypotheses for equality of similarity and for equality of the samples. We can then search for biological or environmental reasons for the dissimilarities.

If we discover that similarities are not random, we can investigate whether the pattern of species presences in relation to environmental variables is non-random. The measure of matrix disorder as proposed by Atmar and Patterson (1993) calculates the entropy of matrix as measured by temperature. Temperature measures the deviation from complete order (0°) to complete disorder (100°) in which the cells of a matrix are analogous to the positions of gas molecules in a rectangular container. After the container has been maximally packed to fill the upper left corner (by convention), the distribution of empty and filled cells determines the degree of disorder in the species distributions and corresponds to a temperature that would produce the degree of disorder (Atmar and Patterson, 1993). To test whether the temperature could be obtained due to random effects, 500 Monte Carlo simulations of randomly determined matrices of the same geometry were calculated. The significance of observed temperatures is ascertained in relation to the variance of the simulated distributions. Two patterns are consistent with a rejection of the null hypothesis by this method: that the distributions are nested subsets (Atmar and Patterson, 1993); or there is a clinal turnover in fauna. Software to calculate matrix disorder is available online from Atmar and Patterson: http://www.fieldmuseum.org.

A simulation procedure was used to evaluate the large number of unique species that were found in the Upper Río Paraguay subregion. The overall area of the Río Paraguay was broken into five subregions. Samples from the five subregions contained different number of species, the largest num-

ber was collected in the Upper Paraguay. Because the Upper Paraguay had the largest number of species overall, perhaps we should also expect the Upper Paraguay to contain the largest number of species that were collected only there. This proposition became the null hypothesis and was tested as follows. Five samples were constructed randomly from the total list of 173 fish species that were collected such that each pseudo-subregion sample contained the same number of species as its corresponding subregion sample. The number of unique species was then determined in the pseudo-Upper Paraguay sample. This procedure was repeated 100 times to yield a distribution of the number of unique species found in the pseudo-Upper Paraguay samples (Fig. 7.6). The number of unique species observed in the Upper Paraguay sample was compared to the mean using standard deviations of the distribution based upon random processes.

Results

I. Effort

Analyses of the number of fish species found within subregions, or within macro-habitats, is potentially confounded by the number of collections (field stations) for the subdivisions within each. Confounding effects of differential effort were examined by regressing the number of species collected against the number of collection stations. Collections within the Río Paraguay were divided into five subregions (Map 2). The regression line in the scatter plot (Fig. 7.4) was not significantly different from that with a zero slope (F=5.26, df=1&3, P<0.11). The standard error of the estimate was 22.7. From this we conclude that the number of stations did not significantly affect the number of species captured in a subregion. For example, 57 stations were collected in the Upper Paraguay subregion from which 105 species were recorded. However, roughly half the effort in the Lower

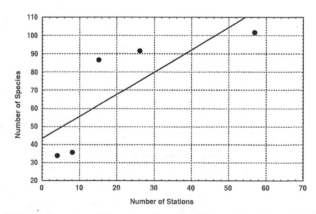

Figure 7.4. Regression of the number of species of fishes on the number of collecting stations for the five subregions within the Río Paraguay.

Paraguay subregion (26 stations) resulted in almost the same number of species (93 species). In the Apa subregion 85 species were collected from only 15 stations.

The situation with macro-habitats is slightly more complex. Ten distinct macro-habitats were sampled and are described below. The scatter plot and regression line are shown in Fig. 7.5a. The unconstrained regression equation, Y=15.62 + 2.74 X, where Y is the number of species and X is the number of stations, is marginally significant. The intercept is not significantly different from zero (t=1.081, df=8, P<0.311) but the slope is significant at the 5% level (F=6.27, df=1&8, P<0.04). However, if we constrain the regression to go through the origin (0 species, 0 stations), then we fail to reject the null hypothesis of a zero slope (P<0.07).

These results are entirely dependent upon a single outlier represented by the Río Paraguay beach habitat from which 110 species were collected, more than 35% greater than the macro-habitat next richest in number of species – flooded forests with 68. If we treat the Río Paraguay beach sample as an outlier, and recalculate the regression without it (Fig.

7.5b), then again slope is not significantly different from zero (P<0.44) and the R_2=0.09. Removal of no other single point affects the significance of the original regression, which means that the marginal significance is entirely due to the outlier, and the relationship between number of species collected and the number of field stations is not evident in the nine other macro-habitat samples.

We, conclude, therefore, that analyses of species richness among subregions and macro-habitats is little or not at all confounded by the number of sampling stations.

II. Subregions

The section of the Río Paraguay basin surveyed during the expedition was divided into five subregions (Map 2) as follows: Upper Paraguay; Lower Paraguay; Río Apa; Río Negro; and Riacho La Paz. These subregions show important patterns of similarities and differences that will be critical for the formation of a conservation plan for the freshwater fishes. For example, of the 173 species collected among all subregions, only nine species (=5.2%) were found in each of the five subregions (Table 7.1). In contrast, Chernoff et al. (1999) found 15.6% of the 313 species living in all six subregions of the Upper Río Orthon Basin of Bolivia.

The Upper Río Paraguay had the highest species richness of the subregions in which 105 or 60.7% of all species were collected but the Lower Río Paraguay and the Río Apa subregions were not appreciably less diverse (Table 7.2). These three regions account for 94.2% of all the species collected.

A

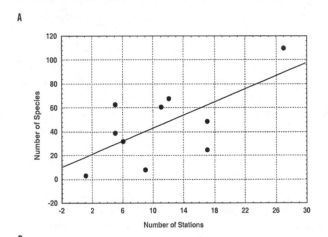

B

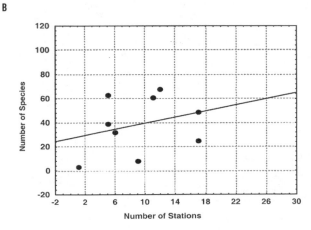

Figure 7.5. Regression of the number of species on the number of collecting stations for the 10 macrohabitats (a) and for nine macrohabitats (b) treating the Rio Paraguay beach sample as an outlier.

Table 7.1. Fish species found living in all subregions (n=9), only in the Riacho La Paz (n=4) and only in the Río Negro (n=6).

All Subregions	Riacho La Paz Only
Aphyocharax anisitsi	*Characidium* sp. 2
Characidium cf. *fasciatum*	*Potamotrygon histrix*
Crenicichla lepidota	*Roeboides* cf. *descalvadensis*
Holoshesthes pequira	*Triportheus* n. sp. A
Moenkhausia dichroura	
Odontostilbe paraguayensis	
Psellogrammus kennedyi	
Pyrrhulina australis	
Steindachnerina brevipinna	

Río Apa and Riacho La Paz Only	Río Negro Only
Astyanax fasciatus	*Cheirodon* cf. *stenodon*
Astyanax lineatus	*Hoplosternum littorale*
Brachychalcinus retrospina	*Lepthoplosternum pectorale*
Bryconamericus sp. 1	*Megalechis thoracata*
Otocinclus Maríae	*Monotocheirodon* cf. sp.
Piabarchus cf. *analis*	*Rhamphichthys rostratus*

The Río Negro and Riacho La Paz accounted for only 10 species not found in the other subregions (Table 7.2). This is not surprising because collections were not made far above the mouths of these tributaries to the Río Paraguay.

What is surprising, however, is the number of species uniquely found in the Upper Río Paraguay, Lower Río Paraguay and Río Apa subregions that range between 19.4–28.2% of their samples (Tables 7.1–7.4). A relatively high percentage would be expected from the Río Apa subregion because it constitutes a very different subregion and set of habitats than are found in the Río Paraguay subregions. While some of the fishes collected only in the Río Apa, such as *Pimelodella laticeps* and *Thoracocharax stellatus*, are probably transients, others such as *Imparfinis minutus*, *Jupiaba acanthogaster*, and *Xenurobrycon macropus*, are not.

The interpretation of the number of unique species in the Río Paraguay subregions is somewhat problematic. Could the relatively high numbers be due to random effects? That is, if community composition was entirely random, how many unique species should we expect in a sample of any given size. This was evaluated by simulating 100, random, five-subregion communities such that the number of species present for each simulated subregion matched exactly the numbers observed: 105, 93, 85, 35, and 41. For the set of random samples of the size of the Upper Río Paraguay subregion (Fig. 7.6) we find a mean of 15.7 unique species ± 3.3 species. The observed value of 27 unique species is 3.4 standard deviations above the mean ($P<0.0002$); significantly more than expected by random processes. The fact that most of the species collected only in the Upper Río Paraguay are known to have much larger distributions may indicate that this region is not necessarily unique but rather influenced by the greater diversity found in the northern portions of the basin, including the Pantanal (Bonetto, 1986).

The numbers of unique species found in the Río Apa and in the Lower Río Paraguay are also greater than one would expect due to random chance given the entire sample of species collected during the expedition. The number of unique species (n=18) collected in the Lower Río Paraguay contains many species that are clearly artifacts such as: *Potamotrygon motoro*, *Serrasalmus elongatus*, *S. spilopleura*,

Table 7.3. Fish species collected only in the Upper (n=27) and Lower (n=18) portions of the Río Paraguay, respectively.

Upper Río Paraguay Only	Lower Río Paraguay Only
Anadoras grypus	*Ageneiosus* cf. *brevifilis*
Ancistrus sp.	*Bryconamericus exodon*
Astronotus crassipinnis	*Charax stenopterus*
Brycon sp.	*Galeocharax gulo*
Chaetobranchopsis australe	*Hemigrammus* cf. *tridens*
Characidium sp. 1	*Hyphessobrycon* sp. 1
Charax sp.	*Hyphessobrycon* sp. 2
Cheirodon sp. 3	*Leporinus* cf. *obtusidens*
Cheirodontinae sp.	*Microcharacidium* sp.
Cichlasoma sp.	*Pimelodus blochii*
Cochliodon sp. 1	*Pimelodus maculatus*
Crenicichla sp.	*Pinirampus pirinampu*
Crenicichla cf. *vittata*	*Potamotrygon motoro*
Doras eigenmanni	*Rivulus* sp.
Doras sp.	*Serrasalmus* cf. *elongatus*
Hemiodus orthonops	*Serrasalmus spilopleura*
Holoshesthes pequira	*Sorubim lima*
Hyphessobrycon maxillaris	
Hypostomus sp. 2	
Lepidosiren paradoxa	
Leporinus friderici	
Loricaria sp. 1	
Megalechis thoracata	
Mylossoma duriventre	
Piaractus mitrei	
Pimelodella sp.	
Salminus maxillosus	

Table 7.4. Fish species collected only in the Río Apa (n=24).

Apteronotus albifrons
Bunocephalus doriae
Callichthys callichthys
Cheirodon sp. 1
Clupeacharax anchoveoides
Corydoras aeneus
Corydoras cf. *ellisae*
Curimatella dorsalis
Jupiaba acanthogaster
Farlowella paraguayensis
Heptapterus sp. (nov.?)
Hypostomus sp. 3
Imparfinis minutus
Leporinus friderici acutidens
Leporinus striatus
Odontostilbe sp.
Pimelodella laticeps
Poptella paraguayensis
Pseudocetopsis gobioides
Pseudohemiodon laticeps
Roeboides prognathus
Tatia aulopygia
Thoracocharax stellatus
Xenurobrycon macropus

Ageneiosus brevifilis, Pimelodus blochii, P. maculatus, Pinirampus pirinampu, and *Sorubim lima.* If we subtract minimally these seven species from the list, then the number of unique species within the Lower Río Paraguay is not significantly different (P<0.22) from that expected by random cause. Though we will conclude (see below) that there is an important non-random relationship between these subregions, the differences due to the number of unique species reflect the differences in macro-habitats that were sampled within each of these two subregions.

The similarities among the subregions are shown in Table 7.5. In all but three of the 10 instances, the null hypothesis of random association was rejected; the probability of obtaining the mean index of similarity, S'_s, at random was less than 1×10^{-4}. That the significant similarities (Table 7.5) are above the mean random similarity, S^*_s, signifies that there is biological dependence among the subregions (Chernoff et al., 1999). The biological dependence may take the form of nested subsets, clinal turnover, etc., that may be due to size (Atmar and Patterson, 1993), hydrological connections or macro-habitats (see below). For example, the collections from the Río Negro have a very high similarity with the Upper Río Paraguay (S'_s > 80%) (Table 7.5). The former shares 29 of its 41 species (Table 7.2) with the latter, into which it flows. Because the collections were taken from the lower reaches of the Río Negro, in a relatively restricted area, this may comprise a nested subset of the larger Upper Río Paraguay subregion. A second example is the Riacho La Paz that has 29 of its 35 collected species also present in the Río Apa (S'_s > 85%) (Tables 7.2,5). In this instance, however, the two rivers are not confluent but rather they are adjacent drainages on the east side of the Río Paraguay. A nested relationship between the two would be due to similar macro-habitats found in these tributaries.

Given that the similarities and, hence, differences among subregions are not random, we used entropy statistics to test if they form an overall pattern. If the distribution of species among all subregions, despite the level of similarity, is random then there will be a high degree of disorder in the matrix and the temperature will tend towards 100° as

entropy increases. The results of the entropy statistics show that the matrix had a temperature of 33.14°. Based on a perfectly ordered matrix of temperature of 0°, the observed temperature is significantly cooler (i.e., has more structure) than that predicted by 500 Monte Carlo simulations (Fig. 7.7) such that the probability of obtaining the observed temperature at random is 1.07×10^{-4}. Thus, the matrix of species distributions exhibits a significant pattern despite the idiosyncratic occurrences of the unique species (Tables 7.1, 3–4).

There are two possibilities to explain the distributional pattern of fishes that we may observe among subregions within the Río Paraguay basin: nested subsets as discussed by Atmar and Patterson (1993) or clinal variation (i.e., smooth turnover). Clinal transitions in fauna should roughly correspond to an isolation by distance model (Sokal and

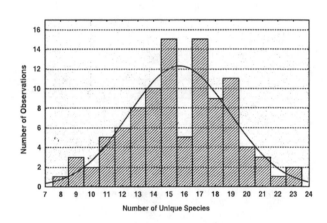

Figure 7.6. Distribution of 100 random simulations of the number of unique species found in the most species rich subregion given the constraints of five subregions with 105, 93, 85, 35 and 41 species respectively, drawn from a total population of 173 species.

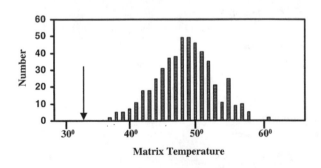

Figure 7.7. Distribution of matrix temperatures based upon 500 Monte Carlo simulations (histogram) of the five subregion by 173 species matrix. The distribution has a mean of 480 and a standard deviation of 4.010. The observed matrix temperature, 33.140, is 3.71 standard deviations below the mean of the random simulations. The probability of obtaining the observed temperature by chance alone is 1.07×10^{-4}.

Table 7.5. Means of Simpson's Index of Similarity, S's, among ichthyofaunas found in five subregions of the Río Paraguay. Larger samples were rarefied 200 times to the size of the smaller samples. Index values are reported as percentages. ** indicates P<0.0001. Abbreviations: APA – Río Apa; LP – Lower Río Paraguay; RLP – Riacho La Paz; RN – Río Negro; UP – Upper Río Paraguay.

	UP	LP	APA	RLP
LP	69.6**			
APA	54.0	52.9		
RLP	58.8**	64.7**	85.3**	
RN	80.6**	72.2**	52.8**	29.4

Oden, 1976) in which high similarity among adjacent or proximate regions results from the sharing of species. The similarity declines smoothly as distance among regions increases; there may be little or no species common to distant endpoints. Both the minimum spanning tree from the distance matrix (1-S'_s) and the Camin-Sokal parsimony analysis from the matrix of species distributions (Appendix 13) produce the identical pattern of relationships among the subregions (Fig. 7.8).

The results show two groupings: (i) Río Negro – Upper Río Paraguay – Lower Río Paraguay; and (ii) Río Apa – Riacho La Paz. The two groups represent a larger, riverine

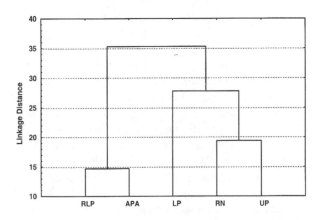

Figure 7.8. Camin-Sokal parsimony cluster analysis of five subregions in the Río Paraguay. Abbreviations: APA–Río Apa; LP–lower Río Paraguay; RLP–Riacho de la Paz; RN–Río Negro; and UP–upper Río Paraguay.

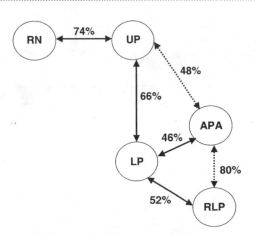

Figure 7.9. Gabriel network among the five subregions (in circles) of the Río Paraguay. Values indicate mean Simpson's Similarity of connected subregions. Solid lines indicate adjacent subregions connected directly by river. Dashed lines indicate comparisons of non-adjacent subregions. Abbreviations: APA–Río Apa; LP–lower Río Paraguay; RLP–Riacho La Paz; RN–Río Negro; and UP–upper Río Paraguay.

fauna and a tributary fauna, respectively, with some overlap (Figs. 7.8–9). The alternative hypothesis of faunal change as a function of distance is not supported by the data. For example, the fishes found in the Río Apa are equally similar to those found in both the Upper and Lower Río Paraguay subregions (Fig. 7.9). Furthermore, the relatively high number (n=18) of fish species found uniquely in the Lower Río Paraguay is largely artifactual—most of those fishes are recorded in the basin above the Upper Río Paraguay subregion in the Pantanal (Toledo-Piza et al., 2001).

The overlap of the Río Apa samples with respect to the Upper and Lower Río Paraguay subregions (i.e., approximately 50% similarity, Fig. 7.9) was estimated as due to random effects (Table 7.5) and not due to biological structuring. In the case of the slightly higher and non-random similarity between the Riacho La Paz and the Lower Río Paraguay, the overlap is due to the fact that samples were not taken very far from the mouth of the riacho. Nonetheless, the Riacho La Paz forms a group with the Río Apa because they share a very high number of taxa, including six species (Table 7.1) not found elsewhere (=14.6% of the Riacho La Paz collection) of which only *Astyanax fasciatus* may be artifactual.

The data also indicate that within each of the two groups, the subregions relate to each other as nested subsets. That is, the larger areas sampled contain more taxa. This important ramifications for conservation plans and in our analyses of threats. Furthermore, sufficient habitat needs to be preserved from each of the riverine and the tributary subregional groups for potential recolonization.

Table 7.6. Species of fishes captured in the trawl (n=8), in the rapids (n=3), only along Río Apa beaches (n=12), and only in backwaters (n=1).

Trawl	Río Apa Beaches Only
Cichlasoma sp.	*Callichthys callichthys*
Holoshesthes pequira	Characidae sp.
Loricaria sp. 1	*Clupeacharax anchoveoides*
Loricaria sp. 2	*Corydoras* cf. *ellisae*
Moenkhausia sanctaefilomenae	*Hypostomus* sp. 3
Pimelodella gracilis	*Jupiaba acanthogaster*
Steindachnerina conspersa	*Leporinus friderici acutidens*
Synbranchus marmoratus	*Leporinus striatus*
	Odontostilbe sp.
Rapids	*Pimelodella laticeps*
Bryconamericus sp. 1	*Roeboides prognathus*
Holoshesthes pequira	*Tatia aulopygia*
Imparfinis minutus	
	Backwaters Only
	Microcharacidium sp.

III. Macro-habitat

We recognized 10 macro-habitats in the Río Paraguay with their abbreviations as follows: Río Paraguay beach (BP); Río Apa beach (BA); backwater (BW); flooded forest (FF); floating vegetation (FV); lagoon (LG); main channel (MC); clear water (CW); rapids (R), and trawl (T). The trawl samples should normally describe river bottom communities but the trawl was not successful, capturing only a total of eight species. A number of the species captured by the trawl are not very indicative of deeper, river bottom communities; i.e., the *Moenkhausia sanctaefilomenae, Holoshesthes pequira,* and the *Cichlasoma* sp. (Table 7.6). The two unidentified *Loricaria* spp. and the *Pimelodella gracilis* are better indicators of the river bottom. However, because the trawl samples had low species richness we did not use the trawl samples in the numerical analyses. Similarly, the sample from the rapids contained only three species (Table 7.6), and were not analyzed further. Therefore, eight macro-habitats were included in the subsequent analysis.

By far the most species rich habitats were the beaches along the Río Paraguay with more than 1.6 times the number of fish species found in the next largest macro-habitat, flooded forests (Table 7.7). Clear water and Río Apa beach macro-habitats join the flooded forests in the second tier of species richness (Table 7.7). As will be discussed below, there is a non-random pattern to species associations among macro-habitats but the pattern is one of large faunal overlaps and dependence.

With the exception of the main channel, large numbers of fish species were found in a number of macro-habitats (Table 7.7). Remarkably few species, however, were found exclusively in two macro-habitats (Table 7.7). Indeed, only four of the eight macro-habitats—BP, BA, MC and CW—had a relatively large number of unique species. In fact, almost 50% of the main channel species were found only in these habitats (Tables 7.6, 7.8, 7.9). While these results imply that species are widely dispersed among many of the macro-habitat types, there are also limitations on species distributions. There were no species that were found to live in more than six macro-habitats. Fifteen and 35 fish species were found in six and five macro-habitats, respectively (Table 7.8). The number of fish species found in fewer and fewer macro-habitats continues to increase exponentially until the number of species collected (n=173) is reached.

The patterns of similarities (Table 7.10) show that for only eight of the 28 comparisons did we fail to reject the null hypotheses of similarity due to randomness, even when the significance levels were corrected by the sequential Bonferroni procedure. For the 21 comparisons in which the null hypothesis were rejected, all had more species in common than the random mean (Table 7.10). These results indicate that there are dependence relations among the macro-habitats.

Of the eight similarities that do not fall outside the predictions of random distributions, six involve comparisons to the main channel (Table 7.10). Only the similarity of the main channel to the Río Paraguay beach habitats was significantly different from random (Table 7.10). The main channel habitat extends into deeper water, off of the beaches and shores. Many of the fish species found in the main channel were slightly larger species (e.g., the dorado, *Salminus maxillosus,* or the sorubí, *Sorubim lima* (Table 7.8)) and were not taken in other habitats. Given the adjacent spatial relationship of the beach and the main channel within the Río Paraguay, it is not surprising that a relatively high percentage

Table 7.7 The number of species shared (lower triangle) and the number of species shared uniquely (upper triangle) among the macro-habitats within the Río Paraguay. Abbreviations: BA–beach Río Apa; BP–beach Río Paraguay; BW–backwaters; CW–clear water; FF–flooded forests; FV–floating vegetation; LG–lagoons; MC–main channels; n–total number of species; u–number of uniques species; %u–percentage of unique species to number of species within a macro-habitat.

	BP	BA	BW	FF	FV	LG	MC	CW
BP		0	2	6	1	0	2	2
BA	41		0	0	0	0	1	5
BW	31	20		0	0	0	0	0
FF	60	28	25		0	0	1	0
FV	45	26	23	37		0	0	0
LG	34	17	18	29	21		0	0
MC	10	6	1	7	3	1		0
CW	44	37	19	34	26	19	3	
n	110	61	32	68	49	39	25	63
u	25	12	1	5	4	4	12	11
%u	22.7	19.7	3.1	7.4	8.2	10.3	48.0	17.4

of the species collected in the main channel should also be collected along the beaches.

In order to determine whether there is an overall pattern to the distribution of the fishes among the macro-habitats, we used the matrix entropy method of Atmar and Patterson (1993). The results of 500 Monte Carlo simulations (Fig. 7.10) show that the matrix of species by macro-habitat distributions is significantly cooler than would be predicted at random. The observed matrix temperature is 17.6°, more

than 10.5 standard deviations below the mean taken from the random permutations—the probability of obtaining the observed matrix temperature at random is 9.3×10^{-26}. Thus, the distributions of species among macro-habitats is much more orderly than would be expected at random. The idiosyncratic distributions of a number of species was not sufficient to disrupt an overall pattern of order. The possible patterns included nested and faunal turnovers.

The overall pattern was sought from two tree analyses, one based upon Camin-Sokal parsimony and the other a minimum spanning tree. Two different topologies of relationships were obtained (Fig. 7.11). The analysis using Camin-Sokal parsimony (Fig. 7.11a) emphasizes the dissimilarities of the main channel macro-habitat from the others. Within the remaining group two clusters are present: (i) macro-habitats primarily associated with the mainstem of the Río Paraguay; and (ii) a cluster containing the Río Apa beach and clear water macro-habitats. The larger cluster is very interesting because it extends from the beaches into the

Table 7.8. Species of fishes found in any five (n=35) or six (n=15) macro-habitats, only in lagoons (n=4), and only in the main channel (n=12).

Five Habitats	Six Habitats
Apareiodon affinis	Aphyocharax anisitsi
Aphyocharax anisitsi	Apistogramma trifasciata
Aphyocharax dentatus	Astyanax paraguayensis
Apistogramma borellii	Cheirodon piaba
Apistogramma commbrae	Holoshesthes pequira
Apistogramma trifasciata	Hyphessobrycon eques
Astyanax paraguayensis	Moenkhausia dichroura
Bryconamericus exodon	Moenkhausia sanctaefilomenae
Characidium cf. fasciatum	Odontostilbe paraguayensis
Cheirodon piaba	Otocinclus vestitus
Cichlasoma dimerus	Otocinclus vittatus
Crenicichla lepidota	Prionobrama paraguayensis
Curimatella dorsalis	Psellogrammus kennedyi
Eigenmannia trilineata	Pyrrhulina australis
Hemigrammus cf. lunatus	Roeboides paranensis
Holoshesthes pequira	
Hoplias malabaricus	**Lagoons Only**
Hyphessobrycon eques	Hemigrammus cf. tridens
Hypoptopoma cf. inexpectata	Rivulus sp.
Hypostomus sp. 1	Serrasalmus cf. elongatus
Moenkhausia dichroura	Serrasalmus spilopleura
Moenkhausia sanctaefilomenae	
Odontostilbe paraguayensis	**Main Channel Only**
Otocinclus vestitus	Ageneiosus cf. brevifilis
Otocinclus vittatus	Auchenipterus nuchalis
Prionobrama paraguayensis	Galeocharax gulo
Psellogrammus kennedyi	Hemisorubim platyrhynchos
Pyrrhulina australis	Hoplosternum littorale
Rineloricaria parva	Pimelodus blochii
Roeboides paranensis	Pimelodus maculatus-blochii
Steindachnerina brevipinna	Pinirampus pirinampu
Steindachnerina conspersa	Potamotrygon motoro
Tetragonopterus argenteus	Rhaphiodon vulpinus
Triportheus nematurus	Salminus maxillosus
Triportheus paranensis	Sorubim lima

Table 7.9. Species of fishes found only in Río Paraguay beaches (n=25), flooded forests (n=5), floating vegetation (n=4), and clear water (n=11).

Río Paraguay Beaches Only	Flooded Forests Only
Apteronotus albifrons	Ancistrus sp.
Astronotus crassipinnis	Hyphessobrycon maxillaris
Bryconamericus sp.	Hypostomus sp. 2
Chaetobranchopsis australe	Leporinus cf. obtusidens
Characidium sp. 1	Piaractus mitrei
Charax sp.	
Cheirodon cf. stenodon	**Floating Vegetation Only**
Cheirodontinae sp.	Anadoras grypus
Cichlasoma sp.	Bryconamericus exodon
Cochliodon sp. 1	Lepthoplosternum pectorale
Crenicichla sp.	Megalechis thoracata
Crenicichla cf. vittata	
Doras sp.	**Clear Waters Only**
Heptapterus sp. (nov.?)	Astyanax fasciatus
Holoshesthes pequira	Astyanax lineatus
Hyphessobrycon sp. 2	Bunocephalus doriae
Lepidosiren paradoxa	Characidium sp. 2
Megalechis thoracata	Charax stenopterus
Monotocheirodon sp.	Cheirodon sp. 1
Pimelodella sp.	Curimatella dorsalis
Pseudocetopsis gobioides	Hyphessobrycon sp. 1
Pseudohemiodon laticeps	Microglanis cf. parahybae
Rhamphichthys rostratus	Potamotrygon histrix
Roeboides cf. descalvadensis	Thoracocharax stellatus
Triportheus n. sp. A	

lagoons—an order that extends from the shores inland. Whether this forms a single nested subset or a pattern of species turnover will be discussed below. The group containing the Río Apa beach and clear water macro-habitats emphasizes the five fish species shared uniquely by them rather than the fact that their highest similarities are to the beaches of the Río Paraguay. These species (e.g., *Brachychalcinus retrospina, Corydoras aeneus,* and *Pimelodella mucosa*), prefer smaller habitats with sandier bottoms and less than turbid waters. All of the clear water habitats were either tributaries to the Río Apa or adjacent to it (e.g., Riacho La Paz).

The minimum spanning tree (Fig. 7.11b) is very similar to the results obtained above. The main channel is still the outlier and the larger cluster containing the Río Paraguay beaches is present but now the clear water and Río Apa beach habitats join sequentially. This large chain emphasizes a pattern of declining similarity to the Río Paraguay beach

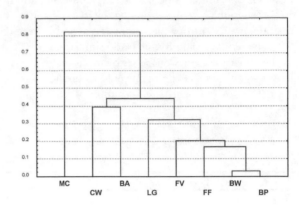

A

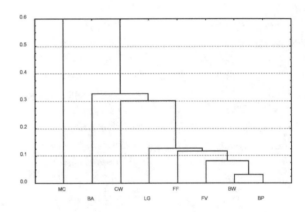

B

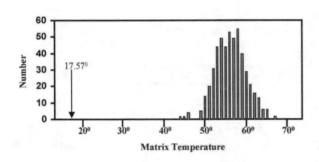

Figure 7.10. Distribution of matrix temperatures based upon 500 Monte Carlo simulations (histogram) of the eight macrohabitat by 173 species matrix. The distribution has a mean of 55.780 and a standard deviation of 3.630. The observed matrix temperature, 17.570, is 10.52 standard deviations below the random mean. The probability of obtaining the observed temperature by chance alone is 9.34x10-26.

Figure 7.11. Camin-Sokal parsimony (a) and minimum spanning tree (b) cluster analyses among eight macrohabitats found within the Río Paraguay. Abbreviations: BA–Río Apa beach; BP–Río Paraguay beach; BW–backwater; CW–clear water; FF–flooded forest; FV–floating vegetation; LG–lagoon; and MC–main channel of the Río Paraguay.

Table 7.10. Means of Simpson's Index of Similarity, S'$_s$, among ichthyofaunas found in eight macro-habitats of the Río Paraguay. Larger samples were rarefied 200 times to the size of the smaller samples. ** indicates P<0.0001. Abbreviations: BA–beach Río Apa, BP–beach Río Paraguay, BW–backwaters, CW–clear water, FF–flooded forests, FV–floating vegetation, LG–lagoons, MC–main channels.

	BP	BA	BW	FF	FV	LG	MC
BA	67.2**						
BW	96.9**	62.5**					
FF	88.2**	45.9	78.1**				
FV	91.8**	53.1**	71.9**	75.5**			
LG	87.2**	43.6**	56.3**	74.4**	53.8**		
MC	40.0**	24.0	4.0	2.8	12.0	4.0	
CW	69.8**	60.7**	59.4**	54.0	53.1**	48.7**	12.0

Rapid Assessment Program

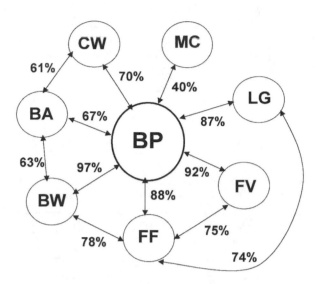

Figure 7.12. Non-hierarchical cluster analysis of eight macrohabitats (in circles) found in the Rio Paraguay basin. Values indicate mean Simpson's Similarity of connected macrohabitats. Abbreviations: BA–Río Apa beach; BP–Río Paraguay beach; BW–backwater; CW–clear water; FF–flooded forest; FV–floating vegetation; LG–lagoon; and MC–main channel of the Río Paraguay.

community. This pattern seems to rule out that the distributions among macro-habitats form nested subsets. That is the fauna present is nested in relation to its size. While the clear water habitats tended to be slightly smaller in size they were not smaller than the backwaters.

Another way to visualize the relationships among the species found in the macro-habitats can be found using a non-hierarchical diagram to represent the significant similarities (Fig. 7.12). The non-hierarchical analysis combines the results from both of the dichotomous tree procedures above. Here, Río Paraguay beach macro-habitat is placed at the center of a concentric arrangement of habitats. In all cases the highest similarities are to the Río Paraguay beaches and then secondarily to other macro-habitats. For example, the backwaters join the flooded forests secondarily, which then join equally but separately to the floating vegetation and to the lagoons. This indicates an apparent use of the environments by fish species at two levels. At the first level, the Río Paraguay beaches serve as a source population for the other macro-habitats. At the second level, fishes are moving among several macro-habitats, for example, from flooded forests into interior lagoons that are interconnected during floods.

Discussion

This chapter has dealt with the distribution of freshwater fishes in a relatively restricted region of the Upper Río Paraguay basin, Paraguay. The basis for the distributional data

came from an AquaRAP survey in September 1997, providing point source information from a rapid assessment. Despite the rapid nature of the survey, we have documented a region that contains a large number of the known fishes of Paraguay including several new records (Toledo-Piza et al., 2001). Yet from a conservation perspective, we have attempted to develop a basis for understanding the dimensions of the distributions of the fishes in the Río Paraguay in order to assess the impact of current or proposed threats.

Threats to the environment of this region (e.g., channelization, logging) will have vastly different impacts depending upon the distribution patterns of the fishes. For example, if all the species found within a region were homogeneously distributed, then environmental destruction to one part of the region or to one type of habitat might not diminish the number of species. However, departure from homogeneity and randomness, whether increasing or decreasing patterns of similarity, could more easily result in loss of species. We first summarize the results above in relation to the two null hypotheses that the distributions of fishes in the Río Paraguay are homogeneous with respect to: (i) subregions and (ii) macro-habitats. We then use that information in a predictive fashion relative to the current environmental threats to the region. We will attempt to provide lists of taxa as profiles for the assemblages that might survive under certain conditions.

I. Distributional Patterns and Null Hypotheses

The results of similarity patterns and their tests against null hypotheses of homogeneity allowed us to reject the null hypotheses for both subregion and macro-habitat effects. For both subregions and macro-habitats, similarities among fish faunas were greater than to be expected at random (P<0.01) indicating interdependency on variables associated with geography or environment. When tested further, we were able to reject (P<<0.001), the null hypotheses of random pattern using measures of disorder in the subregion or macro-habitat by species matrices. In the case of subregions, patterns of similarity, and patterns of shared species, both exclusive and ubiquitous (Figs. 7.8–9), allowed us to discount smooth faunal turnover as the pattern. Rather, the subregion effect is closer to that of two groups, a Río Paraguay group and a Río Apa group, within which there are nested subsets. For macrohabitats, nested patterns are most consistent with a fauna being derived from (i.e., moving through) riverine habitats on seasonal cycles of flooding.

The results are consistent with published literature on the strong geographic and macro-habitat related effects on freshwater fish distributions (Balon and Stewart, 1983; Balon et al., 1986; Hawkes et al., 1986; Mathews, 1986; Ibarra and Stewart, 1989; Cox Fernandes, 1995; Chernoff et al. 1999). Diversity of fish species in the Río Paraguay is a non-random interaction of environmental quality and habitat diversity distributed through the river basin. Only nine out of 173 fish species are found in all subregions; not a single species was in each of the eight major macro-habitats.

The accumulation of species among subregions can best be expressed as one of two faunal groups between which there is a discontinuity (Fig. 7.8). The two regional assemblages are as follows: (i) the Río Paraguay assemblage that includes both upper and lower subregions as well as the Río Negro, and (ii) Río Apa assemblage that includes Río Apa and the Riacho La Paz subregions. Each of the assemblages are characterized by relatively high similarity within assemblage and relatively low similarity among assemblages (Fig. 7.9). We believe that the 66% similarity between the Upper and Lower Río Paraguay subregions is low due to sampling. Many of the species found in the Lower Río Paraguay subregion that were not collected from the Upper Río Paraguay subregion, such as *Hemigrammus tridens, Leporinus obtusidens, Pimelodus blochii, P. maculatus,* and *Sorubim lima,* among others, are known from the Pantanal, just upstream from the Upper Río Paraguay subregion. The Río Apa, Riacho La Paz, and the Río Negro are more characteristic of upland habitat than lowland floodplain river. Even though both the Río Apa and the Riacho La Paz flow into the Lower Río Paraguay, their similarities to it are basically the same as their similarities to the Upper Río Paraguay subregion. The disjunction between the Río Paraguay and the Río Apa assemblages is not as discrete as that among subregions found in the Upper Río Madeira basin (Chernoff et al., 1999). Thirty-five of the 85 species (41%) found in the Río Apa assemblage were not present in the Río Paraguay assemblage (Table 7.11). While a number of these species are collecting artifacts (e.g., *Potamotrygon histrix, Pseudohemiodon laticeps,* and *Triportheus* n. sp. A), the large majority are indicative of smaller, terra-firme tributaries (e.g., *Corydoras* spp., *Imparfinis minutus, Jupiaba acanthogaster, Otocinclus mariae, Piabarchus analis, Xenurobrycon macropus,* etc.). Thus, minimally, there is a 40% faunal shift as one proceeds into these sandier, less muddy tributaries of the Río Paraguay. We predict that more collecting effort above the lower reaches of the Río Apa and the Riacho La Paz will increase the definition of the disjunction.

The Río Paraguay has its headwaters in the northern parts of the basin and flows basically south from the tropics to the temperate zone. While Gorman and Karr (1978) have postulated that the species richness (number of species of fishes) should increase in the lower reaches of rivers, this should not be the case for the Río Paraguay (Bonetto, 1986; Britski et al., 1999). Because the water temperature decreases downstream, the numbers of fish species drops off as one proceeds to the La Plata (Bonetto, 1986). The species richness is much higher in the Pantanal than in the section of the Río Paraguay that we sampled, and there may be a hint of this upstream effect. The Upper Río Paraguay subregion had slightly more species than other regions. Importantly, many of the species collected uniquely in the Upper Río Paraguay

subregion are known from the Pantanal (Table 7.3)(Britski et al., 1999; Toledo-Piza, 2001). The effect may be artifactual. The up- and downstream changes in fish species richness for tropical rivers is not easily documented (Ibarra and Stewart, 1989; Chernoff et al., 1999) and may not be the case.

The pattern of species associations among the macro-habitats was found to be congruent with that for the subregions. The two tree-based analyses (Fig. 7.11) emphasize four

Table 7.11. Fishes collected in the Río Apa and the Riacho La Paz, but not in the Río Paraguay or Río Negro (n=35), and in flooded forests and backwaters.

Río Apa and Riacho La Paz	Flooded Forests and Backwaters
Apteronotus albifrons	*Aphyocharax anisitsi*
Astyanax fasciatus	*Aphyocharax dentatus*
Astyanax lineatus	*Apistogramma borellii*
Brachychalcinus retrospina	*Apistogramma commbrae*
Bryconamericus sp. 1	*Apistogramma trifasciata*
Bunocephalus doriae	*Astyanax paraguayensis*
Callichthys callichthys	*Cheirodon piaba*
Characidae sp.	*Ctenobrycon pelegrini*
Characidium sp. 2	*Holoshesthes pequira*
Cheirodon sp. 1	*Hyphessobrycon eques*
Clupeacharax anchoveoides	*Hypoptopoma* cf. *inexspectata*
Corydoras aeneus	*Hypostomus* sp. 1
Corydoras cf. *ellisae*	*Moenkhausia dichroura*
Curimatella dorsalis	*Moenkhausia sanctaefilomenae*
Jupiaba acanthogaster	*Odontostilbe paraguayensis*
Farlowella paraguayensis	*Otocinclus vestitus*
Heptapterus sp. (nov.?)	*Otocinclus vittatus*
Hypostomus sp. 3	*Parauchenipterus* cf. *galeatus*
Imparfinis minutus	*Potamorrhaphis eigenmanni*
Leporinus friderici acutidens	*Prionobrama paraguayensis*
Leporinus striatus	*Psellogrammus kennedyi*
Odontostilbe sp.	*Pyrrhulina australis*
Otocinclus Mariae	*Rineloricaria parva*
Piabarchus cf. *analis*	*Triportheus nematurus*
Pimelodella laticeps	*Triportheus paranensis*
Poptella paraguayensis	
Potamotrygon histrix	
Pseudocetopsis gobioides	
Pseudohemiodon laticeps	
Roeboides cf. *descalvadensis*	
Roeboides prognathus	
Tatia aulopygia	
Thoracocharax stellatus	
Triportheus n. sp. A	
Xenurobrycon macropus	

important features that are more easily determined from the non-hierarchical cluster analysis (Fig. 7.12).

(I) The most critical feature is that the Río Paraguay beaches are the primary population source for all of the other macro-habitats. The Río Paraguay beaches share the highest number of species with each of the other macro-habitats (Table 7.7). This does not mean that there are not important differences, but rather that the Río Paraguay beaches are the most diverse habitat, with the largest number of species found in all other macro-habitats. All of the similarities to the Río Paraguay beaches are significantly different and higher than the mean of the distribution of random similarity. Hence, there is a dependence of all macro-habitats to the Río Paraguay beach populations as a source. This will be discussed below in the context of recolonization in relation to potential threats.

(II) The patterns of similarities and shared taxa demonstrate the role of floods in structuring faunal assemblages. Waters flood from the main river channel over beaches into the forests and into local depressions forming lagoons, which are then stranded in the dry season. This is reflected in the collected species such that 29 of the 39 species found in lagoons were also found in the flooded forests (Table 7.7); none were found exclusively in these two macro-habitats. Similarly, backwater habitats are formed as water rises or recedes and sits in arms or depressions behind beaches or in bays, until communicating with the main river again. Thus, it is not surprising that the second largest pattern of similarity is between the backwater and flooded forest macro-habitats (Table 7.11). These places accumulate species that prefer quieter habitats and vegetation (e.g., *Apistogramma* spp., *Cheirodon piaba*, *Hyphessobrycon eques*, *Odontostilbe paraguayensis*, *Otocinclus vestitus*, *Pyrrhulina australis*). These species not only have important roles within the ecosystem (e.g., see Lowe-McConnell, 1987), but also have high value in the ornamental fish trade. The role of flooding and macro-habitats for economically important species will be discussed below.

(III) The fish species that inhabit the deeper environments of the main channel (Tables 7.6, 7.8) often come out of the deeper waters to inhabit the beach habitats during the evenings or early mornings. The fish species collected in the trawls and in the gill nets are only a small portion of what is most certainly a much larger fauna (Chernoff et al., 1999). The only significant similarity found between the main channel and any other macro-habitat was with the Río Paraguay beaches at approximately 40%. Had there been more night collections, the similarity might actually have been higher. The point to be made here is that that the deeper water communities differ significantly from shallower habitats (Lowe-McConnell, 1987; Cox-Fernandez, 1995; Chernoff et al., 1999). The linkage for some, but not all, deeper water taxa is to the beach habitats.

(IV) The relationship between the species of fishes found along Río Apa beaches and clearwater macro-habitats is also significant and represents another grouping of the macro-habitats. The Río Apa beaches and clearwater habitats shared 37 species of which five were unique. Both Río Apa beaches and the clear waters were found to share more species with the beaches in the Río Paraguay, and the five uniquely shared

Table 7.12. Fishes species found in both Río Apa beaches and in clear water macro-habitats (n=37).

Ancistrus cf. *piriformis*
Apareiodon affinis
Aphyocharax anisitsi
Aphyocharax dentatus
Apistogramma trifasciata
Astyanax paraguayensis
Brachychalcinus retrospina
Bryconamericus sp. 1
Bryconamericus sp. 2
Bujurquina vittata
Characidium cf. *fasciatum*
Cheirodon piaba
Corydoras aeneus
Crenicichla lepidota
Hemigrammus cf. *lunatus*
Holoshesthes pequira
Hoplias malabaricus
Hyphessobrycon eques
Loricariichthys cf. *platymetopon*
Moenkhausia dichroura
Moenkhausia intermedia
Moenkhausia sanctaefilomenae
Odontostilbe paraguayensis
Otocinclus vestitus
Otocinclus vittatus
Piabarchus cf. *analis*
Pimelodella mucosa
Poptella paraguayensis
Prionobrama paraguayensis
Psellogrammus kennedyi
Rineloricaria lanceolata
Roeboides paranensis
Steindachnerina brevipinna
Steindachnerina conspersa
Steindachnerina insculpta
Tetragonopterus argenteus
Xenurobrycon macropus

species are largely artifactual (e.g., *Steindachnerina insculpta, Poptella paraguayensis*). Nonetheless, the species that are found in these two macro-habitats (Table 7.12) comprise one of the largest assemblages of ornamental and relatively smaller species (<120 millimeters standard length), such as those in the following genera: *Astyanax, Ancistrus, Apareiodon, Apistogramma, Bujurquina, Characidium, Cheirodon, Corydoras, Hemigrammus, Holoshesthes, Moenkhausia, Otocinclus, Odontostilbe*, and *Xenurobrycon*. The environments that provide critical habitat for many of these smaller species beyond the flood zone of the Río Paraguay must become a key element of any conservation plan for the region.

The analyses of both subregions and macro-habitats are congruent. The beach habitats of the Río Paraguay correspond to much of the habitat in the Upper and Lower Río Paraguay subregions. The Río Apa – Riacho La Paz grouping contains most of the Río Apa beach and clearwater habitats. Thus, despite the complexity of some of the patterns that were discovered, we can simplify a preliminary plan for the conservation of the ichthyofauna. In addition, the congruence simplifies the analysis of environmental threats.

II. Analysis of Threats

The major threats facing the Río Paraguay drainage are diverse, with short- and long-term consequences. The threats range from unregulated fisheries to habitat conversion to channelization and dredging. These activities threaten greatly the aquatic ecosystems of the Río Paraguay drainage. Large-scale alterations to the Río Paraguay drainage will result not only in a major loss of biodiversity, but also important commercial losses in fisheries.

In order to establish a baseline for estimation of the biological consequences of environmental threats, we provide a hypothesis about those fish species that might be the most resilient to environmental destruction. In other words, what the fish communities might minimally contain if current threats were implemented to maximum extent. The philosophy for establishing such lists is based upon breadth of fish species distributions across the five subregions and across the 10 macro-habitats that we sampled within the Río Paraguay basin. The breadth of distribution may be taken as an estimator of environmental tolerance or survivability. We define "common taxa" to be those fish species that were found in at least three subregions and in five macro-habitats. There are 31 such species (Table 7.13), which are almost identical to those species found in five habitats (Table 7.8). The list of common species includes taxa from each of the major groups: tetras and their allies (characiforms), catfishes (siluriforms), electric fishes (gymnotiforms), and cichlids (Cichlidae). This assemblage is the one we consider most likely to exist in highly modified habitats but comprises only 17.9% of the species that we captured (Appendix 13), and contains species that depend upon the flood zone.

The following discussion of threats will be divided into four discrete subsections for clarity.

Hidrovia Paraguay-Paraná

The Hidrovia Project was described briefly in the Executive Summary (pp. 17–26). The project intends to channel and, perhaps, straighten the Río Paraguay, and then connect into an Amazon Basin tributary. The immediate threats to the biotic systems come from channelization and river straightening activities. Secondary threats include pollution (bilge water and petroleum products) and interchange of Amazonian and Río Paraguay faunas.

Table 7.13. Common species of fishes found in the Río Paraguay basin (n=31). Common species are those that were collected in three subregions and five macro-habitats.

Apareiodon affinis
Aphyocharax anisitsi
Aphyocharax dentatus
Apistogramma borellii
Apistogramma commbrae
Apistogramma trifasciata
Astyanax paraguayensis
Bryconamericus exodon
Characidium cf. *fasciatum*
Cheirodon piaba
Cichlasoma dimerus
Crenicichla lepidota
Curimatella dorsalis
Eigenmannia trilineata
Hemigrammus cf. *lunatus*
Holoshesthes pequira
Hoplias malabaricus
Hyphessobrycon eques
Hypoptopoma cf. *inexspectata*
Hypostomus sp. 1
Moenkhausia dichroura
Moenkhausia sanctaefilomenae
Odontostilbe paraguayensis
Otocinclus vestitus
Otocinclus vittatus
Prionobrama paraguayensis
Psellogrammus kennedyi
Pyrrhulina australis
Steindachnerina brevipinna
Steindachnerina conspersa
Tetragonopterus argenteus

The Río Paraguay is a relatively shallow river above Asunción and meanders greatly. The depth of the river will need to be increased substantially in order to accommodate ocean going container ships. The installation and maintenance of a commercial shipping channel will require constant dredging (as is done in the Río Orinoco of Venezuela). Rock outcroppings, such as near Cerritos Pão de Açúcar, will have to be eliminated. The overall effect of placing a large channel and straightening of the river will be to change the patterns of hydrography and sedimentation in the river. The deeper channel will actively pull water out of the tributaries, the Chaco, and the Pantanal—the effect is much like expanding the width of a drainpipe in a bathtub. More water can exit the system in shorter time. This will significantly decrease the degree to which the broad flood plains of the Chaco and the Pantanal are inundated during the rainy seasons, and will also decrease the time that flood waters remain out of the main channel. The degree of flooding and the period of inundation will be further decreased as dyking or river straightening is implemented.

Reduction or prevention of flooding in tropical rivers is perhaps the greatest threat not only to biodiversity but also to productivity (Goulding, 1981; Petts, 1985; Lowe-McConnell, 1987; Goulding et al., 1988; Machado-Allison, 1994; McCulley, 1996). Much of the dependence of the fish species in the Río Paraguay on annual flooding is evident in our data. As stated above, one of the major patterns of distributional dependence was that, due to the flooding cycle (Fig. 7.12), more inland habitats shared the majority of their species with both the beach habitats and other inland habitats. Reduction of the inland macro-habitats would remove a critical resource for fishes both for food gathering and for nursery areas. Although beach habitats had the largest number of species (Table 7.7), the number of species found in habitats that result from inundations—backwaters, floating vegetation, flooded forests and lagoons—are almost as great. Ninety species or 52% of the captured species were found in these four macro-habitats. The 90 species comprise both ornamental and commercially valuable food fishes. For example, all of the pacus and pacu-relatives are important components of those habitats including, *Metynnis mola, Myleus tiete, Piaractus mitrei,* and *Mylossoma duriventre.* Other food or potential food fishes included: *Psectrogaster curviventris, Pygocentrus nattereri, Liposarcus anisitsi, Cochliodon* sp. 2, and *Triportheus nematurus.* There were numerous ornamentals that are common in the aquarium trade including, *Gymnotus carapo, Eigenmannia trilineatus, Pyrrhulina australis, Otocinclus* spp., *Hyphessobrycon eques, Gymnocorymbus ternetzi, Bujurquina vittata,* and *Apistogramma* spp.

Seventy-four of the 90 fish species were also found broadly distributed on beaches, again indicating the flood-cycle relationship among the use of macro-habitats by the fishes. Although some of the 16 species not captured on beaches are artifactual (e.g., *Piaractus*), the majority are quiet-water

forms, such as *Hyphessobrycon maxillaris, Hemigrammus tridens, Microcharacidium* sp., and *Rivulus* sp.

Although the "common species" are all found in the list of 90, it is potentially dangerous that these fishes are primarily dependent upon maintenance of the flooding cycle to undergo their life cycles. Many if not most of the species in the list of 90, begin to yolk their eggs and spawn either on the rising floods or directly in the flooded zone (Goulding, 1981; Lowe-McConnell, 1987). During the period of inundation the fishes grow or add fats and leave the inundated areas when the waters recede back into the main channel. Thus, reduction or elimination of the areas of inundation could result in almost 50% loss of the total ichthyofauna. More than half of the 90 species were not discovered in the Río Apa or clear water habitats.

Another consequence of the reduction of waters lateral to the Río Paraguay would be a vast reduction of the Río Apa and Riacho La Paz during the dry season. The Río Apa – Riacho La Paz region contains almost 30 fish species (Table 7.11) that are associated with the terra-firme, headwater-like habitats and were not found elsewhere which are also potentially imperiled by the Hidrovia Project.

For both the Río Paraguay macro-habitats and the Río Apa – Riacho La Paz zone, the Pantanal of Mato Grosso do Sul would be a likely source from which recolonization of the fish fauna might be possible. However, we predict that the Pantanal will shrink in size as well and will suffer heavy casualties. Using the fish sample from this AquaRAP, and the heterogeneous patterns of distribution and the dependency upon the flooding cycle, we calculate that more than 60% of the fish species could be eliminated from the basin if the Hidrovia Project is implemented. The economic costs of this extinction must be explored.

Land Conversion in the Upper Río Paraguay and Río Negro
Currently, the human population along the western side of the Río Paraguay, especially in the upper section and along the Río Negro is low. However, populations are growing, bringing increasing pressures from agriculture and cattle-ranching. The effect is increased deforestation and land-clearing for agriculture with some indications of plans for dyking to control seasonal flood waters. The consequences for land conversion and flood control in the upper section of the Río Paraguay could be a large loss of aquatic biodiversity due to loss of critical nursery and recruitment areas for fishes as well as important macro-habitats for adult fishes.

The upper section of the Río Paraguay and the Río Negro contain no less than the 111 fish species captured in those subregions, a value that represents 64.2% of the total captured fauna. The potential loss of these species is argued in a fashion similar to that presented above for the Hidrovia Project. However, in such a limited case, source populations for recolonization exist in the Pantanal and in the Lower Río Paraguay.

Land Conversion in the Río Apa and the Riacho La Paz

Much of the Río Apa sub-basin is used for agriculture and ranching. The area however, is not stable and continued habitat conversion is underway. Because this section is more arid, the drier soils make it necessary to use river water for irrigation. Additionally, because the waters are clear and not turbid, sedimentation and erosion damage the ecosystem. Much of the Río Apa basin is similar to terra firme and headwater habitats. The system is very delicate.

Along with the Riacho La Paz, there are approximately 30 fish species that are not found in the more lowland habitats. This represents approximately 17.3% of the fauna that is not easily replaced. The closest sources for recolonization are in the headwaters of the Pantanal or in the Río Paraná. These are long distances and we doubt that these fishes are active long-distance colonizers (e.g., *Microcharacidium*). Thus, continued habitat conversion could eliminate these species that are found nowhere else in this portion of the Río Paraguay basin.

The dorado sports-fishery

The Río Paraguay provides fisherman in Brasil and in Paraguay with wonderful sportsfishing for dorado, *Salminus maxillosus* (Bonetto, 1986). Currently, the fish seem abundant, but there are little or no limitations on this fishery. The dorado is a top piscivore and exceeds 30 kilograms in size (Bonetto, 1986). The life history parameters, especially the number of years before individuals reach maturity, are not known for this species. This species is of immediate concern and must be studied to protect this magnificent resource.

Conclusions and Recommendations

The role of rapid assessment programs is to provide a brief but critical snapshot of the composition and structure of ecosystems. Because of the imminent threats to ecosystems and their organisms across the globe (Myers, 1988, 1990; Sisk et al., 1994), a strategy of rapid assessments across landscapes can target the best and most valuable habitats for immediate conservation and detailed study. To maximize the value of the process, the rapid assessments should provide not only surveys of the organisms and vital environmental correlates, but also hypotheses about the community structure within the region, as well as resilience to environmental threats.

This task is particularly daunting in application to aquatic ecosystems within tropical South America because the dynamics and complexity within these ecosystems are often unappreciated (e.g., Goulding 1980, 1981; Goulding et al., 1988; Machado-Allison, 1994; Barthem and Goulding, 1997). The perspectives presented on the importance of and challenges facing conservation of aquatic ecosystems (Naiman et al., 1995; Pringle, 1997), are often based exclu-

sively upon relatively simple temperate systems. Human development strategies (e.g., Almeida and Campari, 1995) within the basins of the principal rivers of South America often completely ignore the impacts to aquatic ecosystems.

The purpose of this paper has been to evaluate the patterns of similarities for the distributions of fish species among subregions and macro-habitats within the section of the Río Paraguay basin between Concepción and the Brazilian border, near the Río Negro. This area was discovered to be fairly rich in fish species with 173 species captured (Toledo-Piza et al., 2001). This is an underestimate of the species richness because the number of additional species being captured on a daily basis had not diminished (Toledo-Piza et al., 2001: Fig. 6.1). A preliminary examination of the structure of the fishes found within the region is critical as a primary basis for designing a conservation strategy in relation to current and impending environmental threats.

The methods we used are designed to test the null hypotheses of homogeneity of distribution of the fishes (or any organisms) in relation to geographic or environmental variables. Non-homogeneous, patterned distributions require different conservation strategies than does the situation where organisms are homogeneously distributed (e.g., Atmar and Patterson, 1993; Christensen, 1997). Our methods are appropriately conservative and robust, and have the advantage in that we can determine whether the similarity is greater than or less than those expected at random.

The use of these methods follows from the use by Chernoff et al. (1999) in order to estimate patterning within an assemblage of species. This in itself is not a conservative approach to community ecology, but we believe that rapid assessment programs must provide beginning insights into conservation plans to protect ecosystems and biodiversity to the maximum extent possible. We are persuaded by the elegant argumentation of Shrader-Frechette and McCoy (1993). The ethics of conservation require that we risk a type-I error (rejection of a true null hypothesis) rather than a type-II error (failure to reject a false null hypothesis). In the case presented here, it is better to err on the side of recognizing structure within a biological assemblage. The conservation plan is more stringent than that required to conserve a homogeneously distributed community (Barrett and Barrett, 1997; Christensen, 1997). We, therefore, conclude and recommend the following:

1. The freshwater fishes are non-randomly distributed with respect to subregions. There is a relatively sharp boundary between the Río Apa – Riacho La Paz and the Río Paraguay with more than a 50% turnover in species of fishes. Therefore, it is critical to designate two areas within the Río Paraguay basin as core conservation areas: 1) the Upper Río Paraguay – Río Negro and 2) the Río Apa – Riacho La Paz. Together they represent more than 80% of the diversity within the region.

2. The Río Apa – Riacho La Paz zone is highly threatened because land conversion is actively changing the characteristic of the ecosystem. Extinction or reduction of the fauna in these streams will have serious consequences for the overall biodiversity because potential sources of organisms for recolonization are very far away (Pantanal headwaters and Río Paraná).

3. The freshwater fishes are non-randomly distributed with respect to macro-habitats within the Río Paraguay basin. Beach habitats serve as source populations for other habitats that extend into the forest. These habitats are interconnected during periods of inundation. Therefore, it is critical that the natural flood cycle be maintained and that broad areas for inundation be modified as little as possible. Interference or serious reduction of the flood cycle and inundated habitats could result in as much as a 60% loss of the species in the basin.

4. In its proposed state, the Hidrovia Project poses a major threat to the aquatic biodiversity of the Río Paraguay basin. The reduction of fishes in the region will have important socio-economic consequences (e.g., loss or reduction of fisheries) in addition to the intrinsic value of biodiversity. If the Hidrovia Project is implemented, then plans should be made ahead of time to attempt to find adequate refugia within which the majority of the biodiversity can be self-supporting.

5. Studies of the biology and fisheries of the dorado, *Salminus maxillosus,* should be undertaken immediately. This should be done jointly between Paraguay and Brasil in a study of the entire Río Paraguay basin, including the Pantanal. Fishing limits on size and season should be established to allow this top predator to be fished in a sustainable manner.

Literature Cited

Almeida, A.L.O. de and J.S. Campari. 1995. Sustainable development of the Brazilian Amazon. New York: Oxford University Press. 189pp.

Atmar, W. and B.D. Patterson. 1993. The measure of order and disorder in the distribution of species found in fragmented habitat. Oecologia, 96:373–382.

Balon, E.K., S.S. Crawford, and A. Lelek. 1986. Fish communities of the upper Danube River (Germany, Austria) prior to the new Rhein-Main-Donau connection. Environmental Biology of Fishes, 15:243–271.

Balon, E.K. and D.J. Stewart. 1983. Fish assemblages in a river with unusual gradient (Luongo, Africa—Zaire system), reflections on river zonation and description of another new species. Environmental Biology of Fishes, 9:225–252.

Barrett, N.E. and J.P. Barrett. 1997. Reserve design and the new conservation theory. *In* Pickett, S.T.A., R.S. Ostfield, M. Schachak, and G.E. Likens (eds.). The ecological basis of conservation: heterogeneity, ecosystems, and biodiversity. Chapman and Hall, New York. Pp. 236–251.

Barthem, R. and M. Goulding. 1997. The catfish connection: ecology, migration, and conservation of Amazon predators. New York: Columbia University Press. 144pp.

Bonetto, A.A. 1986. Fish of the Paraná system. *In* Davies, B.R. and K.F. Walker (eds.). The ecology of river systems. Dordrecht, Netherlands: Dr. W. Junk Publishers. Pp. 573–588.

Britski, H. A., K. Z. S. de Silimon, and B. S. Lopes. 1999. Peixes do Pantanal: manual de identificação. Brasília, Brasil: Embrapa. 184pp.

Chernoff, B., P.W. Willink, J. Sarmiento, A. Machado-Allison, N. Menezes and H. Ortega. 1999. Geographic and macro-habitat partitioning of fishes in Tahuamanu-Manuripi region, Upper Rio Orthon basin, Bolivia: conservation recommendations. *In* Chernoff, B. and P.W. Willink (eds.). A biological assessment of aquatic ecosystems of the Upper Rio Orthon basin, Pando, Bolivia. Bulletin of Biological Assessment 15, Conservation International, Washington D.C. Pp. 51–68.

Christensen, N.L., Jr. 1997. Managing for heterogeneity and complexity on dynamic landscapes. *In* Pickett, S.T.A., R.S. Ostfield, M. Schachak, and G.E. Likens (eds.). The ecological basis of conservation: heterogeneity, ecosystems, and biodiversity. New York: Chapman and Hall. Pp. 167–186.

Cox Fernandes, C. 1995. Diversity, distribution and community structure of electric fishes (Gymnotiformes) in the channels of the Amazon River system, Brasil. Unpublished Ph.D. Dissertation. Durham: Duke University. 394pp.

Goulding, M. 1980. The fishes and the forests: explorations in Amazonian natural history. Los Angeles: University of California Press. 280pp.

Goulding, M. 1981. Man and fisheries on an Amazon frontier. Boston: Dr. W. Junk Publishers. 137pp.

Goulding, M., M.L. Carvalho, and E.G. Ferreira. 1988. Rio Negro: rich life in poor water: Amazonian diversity and foodchain ecology as seen through fish communities. The Hague, Netherlands: SPB Academic Publishing. 200pp.

Gorman, O.T. and J.R. Karr. 1978. Habitat structure and stream fish communities. Ecology, 59:507–515.

Hawkes, C.L., D.L. Miller, and W.G. Layher. 1986. Fish ecoregions of Kansas: stream fish assemblage patterns and associated environmental correlates. Environmental Biology of Fishes, 17:267–279.

Ibarra, M. and D.J. Stewart. 1989. Longitudinal zonation of sandy beach fishes in the Napo River basin, eastern Ecuador. Copeia, 1989:364–381.

Lowe-McConnell, R.H. 1987. Ecological studies in tropical fish communities. New York: Cambridge University Press. 382pp.

Machado-Allison, A. 1994. Factors affecting fish communities in the flooded plains of Venezuela. Acta Biologica Venezuelica, 15(2):59–74.

Matthews, W. J. 1986. Fish faunal 'breaks' and stream order in the eastern and central United States. Environmental Biology of Fishes, 17:81–92.

McCulley, P. 1996. Silenced rivers. New Jersey: Zed Books. 320pp.

Myers, N. 1988. Threatened biotas: "hotspots" in tropical forests. The Environmentalist, 8:1–20.

Myers, N. 1990. The biodiversity challenge: expanded hotspots analysis. The Environmentalist, 10:243–256.

Naiman, R.J., J.J. Magnusin, D.M. McKnight and J.A. Stanford. 1995. The Freshwater imperative. Washington, D.C.: Island Press. 165pp.

Petts, G.E. 1985. Impounded rivers. New York: J.S. Wiley and Sons. 344pp.

Pringle, C. M. 1997. Expanding scientific research programs to address conservation challenges in freshwater ecosystems. In Pickett, S.T.A., R.S. Ostfield, M. Schachak, and G.E. Likens (eds.). The ecological basis of conservation: heterogeneity, ecosystems, and biodiversity. New York: Chapman and Hall. Pp. 305–319.

Rice, W.R. 1989. Analyzing tables of statistical tests. Evolution, 43:223–225.

Rohlf, F.J. and R.R. Sokal. 1995. Statistical tables. 3rd Edition. New York: W.H. Freeman and Co. 199pp.

Shrader-Frechette, K.S. and E.D. McCoy. 1993. Method in ecology. New York: Cambridge University Press. 328pp.

Sisk, T.D., A.E. Launer, K.R. Switky, and P.R. Ehrlich. 1994. Identifying extinction threats. BioScience, 44:592–604.

Sokal, R.R. and N.L. Oden. 1976. Spatial correlation in biology: I. methodology. Biological Journal of the Linnean Society, 10:199–228.

Sokal, R.R. and F.J. Rohlf. 1997. Biometry. 3rd Edition. New York: W.H. Freeman and Co. 887pp.

Toledo-Piza, M., B. Chernoff, D. Mandelburger, M. Medina, J. Sarmiento, and P.W. Willink. 2001. Diversity and abundance of fishes in the Upper and Lower Río Paraguay basin and the Río Apa sub-basin. In Chernoff, B., P.W. Willink, and J.R. Montambault (eds.). A biological assessment of the aquatic ecosystems of the Río Paraguay basin, Departamento Alto Paraguay, Paraguay. Pp. 73–79. RAP Bulletin of Biological Assessment 19. Washington, DC: Conservation International.

CHAPTER 8

Congruence of Diversity Patterns among Fishes, Invertebrates, and Aquatic Plants within the Río Paraguay Basin, Paraguay

Barry Chernoff, Philip W. Willink, Antonio Machado-Allison, María Fátima Mereles, Célio Magalhães, Francisco Antonio R. Barbosa, Marcos Callisto Faria Pereira, and Mônica Toledo-Piza

Abstract

Null hypotheses concerning random distributions of species with respect to subregions and macro-habitats within the Río Paraguay are tested with data from 131 species of macro-crustaceans and benthic invertebrates and 186 species of aquatic plants. The patterns are compared to the results for the distributions of fishes presented by Chernoff et al. (2001). The invertebrate data demonstrate the identical pattern among subregions as evident in the fish distributions. The results support the recognition of two zones: (i) the Río Paraguay zone containing the Upper and Lower Río Paraguay and the Río Negro; and (ii) the Río Apa zone containing the Río Apa and the Riacho La Paz. For all data sets, the Río Paraguay zone has higher species richness than the Río Apa zone. The boundary between the two zones is abrupt, which is also supported by the plant data. Only 11 of 186 species of plants were found in both zones. There is no congruence of pattern among macro-habitats. Both invertebrate and plant data sets contain many values that are not different from mean random similarities. The fish data set provides an unambiguous pattern of habitats within the Río Paraguay zone tied together in relation to inundation of habitats during the flood cycle. The invertebrate set provides one cluster that is consistent with the general principal. The plant data set demonstrates a relationship among shore and sand habitats, which experience greater currents than other macro-habitats. The plants found in backwater habitats had little similarity to other macro-habitats. Based on these observations, it is concluded that significant habitat within each of zone needs to be preserved to maintain a large portion of the biodiversity.

Introduction

In previous chapters, Mereles (2001), Barbosa et al. (2001), Magalhães (2001) and Toledo-Piza et al. (2001) have discussed general patterns of the distribution of aquatic and riparian plants, plankton and benthos, macro-crustaceans, and fishes in relation to subregions and macro-habitats within the Río Paraguay basin. Chernoff et al. (2001) tested null hypotheses about the homogeneity of fish distributions with respect to subregions and macro-habitats. They rejected both null hypotheses that the fish distributional data were obtained at random. The distributions of fishes indicate that there are two broad subregional associations: (i) the Upper and Lower Río Paraguay and the Río Negro and (ii) the Río Apa and the Riacho La Paz. The patterns of distribution associated with macro-habitats is congruent with the subregional analysis. Within the Río Paraguay proper, the similarity among macro-habitats is a function of the flooding cycle. In the Río Apa and the Riacho La Paz, the association among macro-habitats is associated with terra firme, headwater conditions. These results were then used to construct a conservation plan relative to fishes and to evaluate the potential effects of threats.

In this chapter, we examine the generality of the patterns discovered by Chernoff et al. (2001) by co-examining patterns found in benthos, plankton, macro-crustaceans and "aquatic" plants. We will then use the commonality or distinctiveness of the patterns to develop a conservation plan that would protect the majority of the aquatic biodiversity within the region that we surveyed.

Methods

The data sets used for these analyses are found in Appendices 5,6, 8–11. Two sets of analyses were performed, including one for invertebrates and one for aquatic plants. The invertebrate data set was obtained by combining the macro-crustacean data of Magalhães (2001) with the plankton and benthos data from Barbosa et al. (2001). Together these data sets comprise 131 species of invertebrates for which presence-absence information was available for all species for the five subregions: Río Negro, Upper Río Paraguay, Lower Río Paraguay, Río Apa and Riacho La Paz (Appendices 8-10). This is referred to as the full invertebrate data set.

Presence-absence information was available for 23 species of shrimps, crabs and molluscs for nine macro-habitats: Río Paraguay beaches, Río Apa beaches, backwaters, flooded forests, floating vegetation, lagoons, rapids, cryptic (crevices in rocks, logs, etc.), and clear waters (Appendix 11). Because the latter three macro-habitats had only 5 or fewer species, they were not analyzed. Furthermore, three species (*Macrobrachium borellii*, *M. brasiliense*, and *Sylviocarcinus australis*) were eliminated from the dataset because of ambiguity in assigning their macro-habitat for the Río Apa collections. The analysis of macro-habitats, thus, included 20 species of invertebrates and six macro-habitats—the reduced invertebrate data set.

We will refer to the aquatic and riparian plants as aquatic plants for the purpose of this chapter. The aquatic plant data set (Appendices 5,6) contains 185 species. The data do not allow a full sub-regional analysis but rather only a comparison of the flora of the Río Paraguay (upper and lower subregions) with that of the Río Apa. Presence-absence data were collected for the following macro-habitats: shores, flooded banks, semilotic, swamps, and sandy banks. Based upon the definitions provided by Mereles (2001) we will call flooded banks - backwaters, semilotic - flooded forests, and swamps - lagoons in order to use the same language as that for the fish and the invertebrate data. For the fishes and for invertebrates, beach habitats do not specify the type of soils. Whereas, for aquatic plants the presence of sandy soils is significant. Aquatic plants were collected along and on shorelines whether or not a beach was present. Thus, there is not complete congruence between the aquatic plant and the fish plus invertebrate data sets. However, if we make an assumption that aquatic plant shoreline habitats function in the same way as invertebrates and fishes beach habitats (e.g., the zone between deeper waters and areas exposed seasonally), then we can easily estimate the congruence among the data sets.

The patterns constructed within data sets will be compared by inspection and with matrix correlation. Because the presence or absence of any species is used multiple times in the calculation of similarities in a similarity matrix, the matrix correlation coefficient cannot be tested using the distribution of the product moment correlation. To test the significance of a matrix correlation coefficient, a random permutation test was performed with 10,000 iterations. The random permutation tests rewrites the rows and columns of each matrix in each iteration. The proportion of the simulated correlation coefficients greater than the absolute value of the observed matrix correlation coefficient approximates the probability of obtaining the results at random.

Results

I. Subregions

The invertebrate data base shows that the species richness of invertebrates was not distributed equally among all five subregions (Table 8.1). That the fewest number of species was found in the Riacho La Paz is partly an effect of effort because the fewest collections were taken in that tributary. By far the richest subregion was the Upper Río Paraguay where we collected 71 species, 54.2% of all the invertebrates collected. With the exception of the Riacho La Paz, the other subregions were moderately diverse with the Lower Río Paraguay and the Río Apa being slightly more than 50% richer than the samples from the Río Negro (Table 8.1).

The means of Simpson's Similarity Indices are highly variable among subregions (Table 8.1), ranging from 10% to almost 80% similarity. All indices are significantly different ($P<0.01$) from a random distribution among subregions. The low values (e.g., Riacho La Paz and Río Negro or Upper Río Paraguay and Río Apa) are actually less than that expected at random, indicating that species are actively partitioning the basin into distinctive regions. This would also be consistent with a strong faunal turnover between subregions.

The pattern of similarities plotted on the Gabriel network of subregions (Fig. 8.1) reveals the existence of two subregional zones of high similarity. One among the subregions of

Table 8.1. Mean Simpson's Index of Similiarity, S' among macro-crustaceans and benthic invertebrates found living in five subregions of the Río Paraguay basin. Larger samples were rarefied 200 times to the size of the smaller samples. Index values reported as percentages. Abbreviations: APA–Río Apa, LP–Lower Río Paraguay, RLP–Riacho La Paz, RN–Río Negro, UP–Upper Río Paraguay, n–number of species present, u–number of unique species, %u–percentage of unique species.

	RN	UP	LP	APA	RLP
RN					
UP	77.14				
LP	54.29	64.29			
APA	37.14	40.35	37.50		
RLP	10.00	40.00	30.00	75.00	
n	35	71	56	57	20
u	7	21	15	21	4
%u	20.00	29.58	26.79	36.84	20.00

the Río Paraguay and the Río Negro and the other between the Río Apa and Riacho La Paz. The high similarities within each of these subregional zones is due to shared taxa within the zone and not due to uniquely shared taxa (Table 8.2). For example, notice that the Upper Río Paraguay shares 27 of 35 and 36 of 56 species with the Río Negro and the Lower Río Paraguay, respectively. However, only six species were found in the Río Negro and the Upper Río Paraguay exclusively; and none were shared exclusively between the latter two subregions (Table 8.2). Thus, within the subregional zones there are no boundaries.

Between the Río Paraguay and Río Apa subregional zones there is a strong faunal turnover. This boundary is evident whether it is calculated at the confluence of the Lower Río Paraguay or between the Upper Río Paraguay and the Río

Apa (Fig. 8.1). That is, from the Lower Río Paraguay into the Río Apa there is a turnover of more than 35 species (=62.5%).

The existence of the two groups of subregions results from the cluster analysis using Camin-Sokal parsimony (Fig. 8.2). Notice that the order of joining within the Río Paraguay group reflects the hydrological connections. It is also important to note that the pattern (Figs. 8.1–2) is not due to a particular group of invertebrates, but rather the signal is distributed across phyla and demonstrates the importance of broad taxonomic sampling.

These results are identical to those from the analysis of fish distributions (Chernoff et al., 2001). The identical results are manifest in the highly significant matrix correlation coefficient between the fish and the invertebrate similarity matrices (r=0.923, P<0.001). The results indicate the following for both aquatic invertebrates and fishes: (i) the Río Negro - Río Paraguay zone contains taxa associated with a flood-zone ecosystem; and (ii) the Río Apa - Riacho La Paz zone contains taxa associated with terra firme, headwater habitats. Furthermore, the rate of faunal turnover between the zones is rather sharp. For both the invertebrates and the fishes, there is a 60% turnover between the zones.

Species of plants were recorded from the Río Paraguay (Upper and Lower) and the Río Apa. A total of 186 species were encountered from which 147 were found in the Río Paraguay and 50 in the Río Negro (Appendices 5,6). Of the 50 species in the Río Apa, 39 were not found in habitats along the Río Paraguay. The remaining 11 species that are shared between these regions are displayed in Table 8.3. This would appear to support the results based upon the invertebrate and fishes.

Despite the low number of species shared between these two areas, however, the observed Simpson's similarity

Table 8.2. Numbers of species of macro-crustaceans and benthic invertebrates shared (lower triangle) and shared exclusively (upper triangle) among subregions of the Río Paraguay. Abbreviations: APA–Río Apa, LP–Lower Río Paraguay, RLP–Riacho La Paz, RN–Río Negro, UP–Upper Río Paraguay.

	RN	UP	LP	APA	RLP
		6	0	4	0
UP	27		11	0	0
LP	19	36		4	0
APA	13	23	21		7
RLP	2	8	6	15	

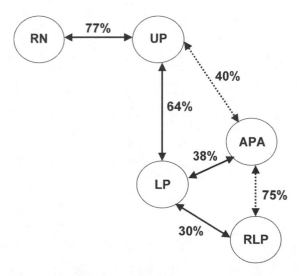

Figure 8.1. Gabriel network summarizing hydrographic relationships within the Río Paraguay. The numbers indicate percent similarity between the connected subregions (in circles). Solid lines indicate a direct river connection. Dashed lines indicate a comparison between subregions not connected directly by river. Abbreviations: APA –Río Apa; LP–lower Río Paraguay; RLP–Riacho de la Paz; RN–Río Negro; and UP–upper Río Paraguay.

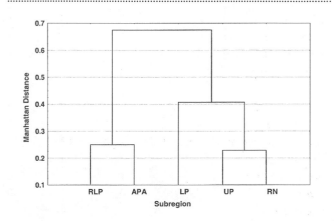

Figure 8.2. Camin-Sokal parsimony cluster analysis of subregions within the Río Paraguay based upon the full invertebrate dataset. Abbreviations: APA – Río Apa; LP–lower Río Paraguay; RLP–Riacho de la Paz; RN–Río Negro; and UP–upper Río Paraguay.

between the two rivers is 22.5%, a value that was not found to differ significantly from the mean random similarity (s=26.5%, std. deviation = 5.38, P>0.05). This indicates that without more information on plant distributions, the low similarity of the Río Apa with respect to the Río Paraguay cannot be interpreted unambiguously. If, however, future studies were to show that the 39 species of Río Apa plants not found in the Río Paraguay were consistent or even larger, then we would conclude that the low similarity is real and not due to the effects of rapid inventory. Nonetheless, the plant information does not contradict the zoological results, it only provides weak or ambiguous support.

II. Macro-habitats

Only 20 of the 131 species of aquatic invertebrates had reliable information about the macro-habitats in which they were collected (Table 8.4). Of the nine macro-habitats that were identified, three contained five or fewer species (rapids, cryptic and clear water) and were not analyzed further. The six remaining habitats were Río Paraguay beach, Río Apa beach, backwaters, flooded forest, floating vegetation and lagoons that contained between 7 and 12 species (Table 8.4). Only six of the 20 species were found in a single macro-habitat, while the number of species shared between habitats varied from four to eight (Table 8.5).

The similarity matrix (Table 8.5) demonstrates a two fold difference in mean similarities from just above 44% to almost 89%. Two hundred simulations for a universe with 20 species total with 7 to 12 species present per macro-habitat indicate that the observed similarities of 50% or less could not be distinguished from random. The remaining

Table 8.3. Aquatic and riparian plant species found along the Río Paraguay and the Río Apa (n=11).

Combretum lanceolatum
Crataeva tapia
Genipa americana
Hydrocotyle ranunculoides
Polygonum punctatum
Salix humboldtiana var. *martiana*
Sapindus saponaria
Senna scabriuscula
Solanum sp.
Triplaris cfr. *guaranitica*
Vitex megapotamica

Table 8.4. Distribution of aquatic invertebrates (n=20) among nine macro-habitats in the Río Paraguay drainage. Abbreviations: BA–Río Apa beach, BP–Río Paraguay beach, BW–backwaters, CR–cryptic, CW–clear waters, FF–flooded forests, FV–floating vegetation, LG–lagoons, RD–rapids.

	BP	BA	BW	FF	FV	LG	RD	CR	CW
Macrobrachium amazonicum	1	1	1	1	1	1	0	0	0
Macrobrachium jelskii	1	0	0	0	0	0	0	0	0
Palaemonetes ivonicus	0	0	0	0	1	1	0	0	0
Pseudopalaemon sp.	0	1	0	0	0	0	0	0	0
Acetes paraguayensis	1	0	0	0	0	1	0	0	0
Dilocarcinus pagei	1	0	1	0	1	1	0	0	0
Poppiana argentiniana	1	0	0	0	0	0	0	0	0
Trichodactylus borellianus	1	1	1	1	1	0	1	1	1
Valdivia camerani	0	0	1	0	1	0	0	0	0
Zilchiopsis oronensis	1	0	0	0	1	0	0	0	0
Gundlachia sp.	1	0	0	1	1	1	0	0	0
Marisa planogyra	0	0	1	0	1	0	0	0	0
Pomacea canaliculata	1	1	1	1	1	1	0	0	0
Pomacea sp.	0	1	0	0	0	0	0	0	0
Biomphalaria sp.	1	1	1	1	1	1	0	0	0
Solaropsis sp.	0	1	0	0	0	0	0	0	0
Omalonyx sp.	0	0	1	1	1	1	0	0	0
Gen. et sp. unindentified	0	1	1	1	0	1	0	0	0
Anodontites sp.	0	1	0	0	0	0	0	1	0
Anodontites crispatus tenebricosus	1	1	0	0	1	0	0	1	0

similarities are significantly different from those expected at random (P<0.01) and reject the null hypotheses.

Camin-Sokal parsimony cannot completely resolve the relationships among the macro-habitats (Fig. 8.3a). The polychotomy is caused by the large number of taxa shared among the flooded forest, floating vegetation, backwater, and lagoon habitats. But apart from the polychotomy, the results of the different clustering methods are congruent (Fig. 8.3). The close relationship between backwater and floating vegetation habitats results from their sharing uniquely two species, *Valdivia camerani* and *Marisa planogyra*.

The major result is that there an intimate relationship among habitats that are seasonally inundated, with the beach habitats being more dissimilar (Fig. 8.3). The group of four inundated habitats share more species in common (6–8) than they do in general with either the Río Apa or Río Paraguay beaches (4–5; Table 8.5). The one exception to this is that eight species were found in common between Río Paraguay beaches and floating vegetation habitats (Tables 8.4, 8.5). This, however, is due to the fact that floating vegetation habitats can extend to the shorelines in proximity to beaches. Nonetheless, the majority of the invertebrate biodiversity is found in less exposed, lentic habitats that are seasonally inundated. The overall pattern based upon shared species or distances emphasizes a Río Paraguay group that communicate vis a vis flooding cycles. The Río Apa beaches are the most distant from the Río Paraguay group.

The pattern of clustering among the macro-habitats, given the invertebrate data, is fairly congruent with the pattern for fishes (compare Fig. 7.11 with Fig. 8.3), in that the inundated habitats form a cluster separated from the Río Apa

beach samples. However, in the case of fishes, the Río Paraguay beaches share the most number of species with backwater habitats. Remember that backwater habitats extend inland and communicate, at least seasonally, with the shorelines or beaches. The differences in branching patterns among invertebrates and fishes is reflected in the relatively low matrix correlation among the sample similarity matrices (Tables 7.10, 8.5; r=0.27, P>0.05) that was not significant. Thus, the similarity between the fishes and the invertebrates is due to the association among habitats that are created during the flood zone along the Río Paraguay. Beach habitats experience the effects of currents, and many species of invertebrates may prefer quieter habitats and may not require access to deeper waters, thereby inverting the association among habitats from that demonstrated for fishes.

A

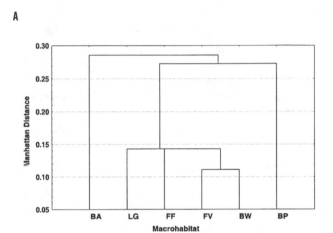

B

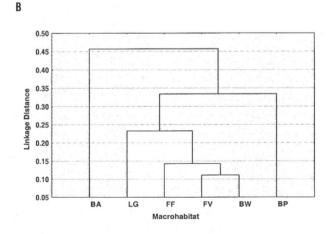

Figure 8.3. Camin-Sokal parsimony (a.) and UPGMA (b.) cluster analyses of macrohabiotats within the Río Paraguay based upon the reduced invertebrate dataset. Abbreviations: BA–beach Río Apa; BP–beach Río Paraguay; BW–backwater; FF–flooded forest; FV–floating Vegetation; and LG – lagoon.

Table 8.5. Number of species shared (upper triangle) and mean Simpson's Index of Similiarity, S', among macro-crustaceans and benthic invertebrates found living in six macro-habitats within the Río Paraguay basin. Larger samples were rarefied 200 times to the size of the smaller samples. Index values reported as percentages. Coefficients shown in bold are significantly different from random (P<0.001). Abbreviations: BA–Río Apa beach, BP– Río Paraguay beach, BW–backwaters, FF–flooded forest, FV–floating vegetation, LG–lagoon, n–number of species present, u– number of unique species, %u–percentage of unique species.

	BP	BA	BW	FF	FV	LG
BP		5	5	5	8	6
BA	50.00		5	5	5	4
BW	55.56	55.56		6	8	6
FF	71.43	71.43	85.71		6	6
FV	72.73	50.00	88.89	85.71		7
LG	66.67	44.44	66.67	85.71	77.78	
n	11	10	9	7	12	9
u	2	4	0	0	0	0
%u	18.18	40.00	0.00	0.00	0.00	0.00

Information about six macro-habitats were collected from all 186 species of plants (Mereles, 2001) (Appendices 5,6). There was more than a doubling of species found among the macro-habitats, and the richest were flooded forests, Río Paraguay shores, and lagoons. Although Mereles (2001) noted that there is usually a high negative correlation between species richness and degree of current, the Río Paraguay shoreline habitats were very rich with 56 species present. The plants exhibited a stronger degree of macro-habitat partitioning than the fishes or invertebrates. For example, the number of unique species (Table 8.6) were exceptionally high ranging from 41% to 86% of the species collected in any habitat. Using the methods of Chernoff et al. (2001), the number of unique species were significantly higher than random expectations (P<0.001). Furthermore, out of 186 species, there were no species found in all five macro-habitats. Only five species (*Pistia stratiotes*, *Crataeva tapia*, *Combretum lanceolatum*, *Polygonum punctatum*, and *Mikania periplocifolia*) were found in four macro-habitats. Fourteen species were found to occupy three macro-habitats.

The matrix of similarity coefficients (Table 8.6) shows that the coefficients range from 0% to 44.4% similarity. However, because of the relatively few taxa found in each macro-habitat out of the large total, the standard deviations for the randomly simulated data are rather high. Only seven of the coefficients are significantly different from random (P<0.001). The remaining eight comparisons cannot be distinguished from a random distribution (Table 8.6). Of the significant coefficients, only two (Ríos Paraguay shores and Apa shores; Río Paraguay shores and sandy habitats) are on the positive tail of the distribution. That is the similarities are greater than expected by chance and illustrate the biological effects of dependence. Thus, even though there is a strong

regional effect between the Río Paraguay and Río Apa subregions, the disparity is not a function of the shore macro-habitats.

Interestingly, the remaining five significant coefficients are on the negative tail of the distribution (i.e., more than three standard deviations below the random mean similarity). These lower than expected similarities indicate strong habitat partitioning among the aquatic plants. Extreme cases are found in backwater or flooded bank habitats such that they possess no species in common with flooded forests (semilotic), lagoons (swamps) or sandy habitats. The flooded forest habitats also show fewer than expected species in common with shore macro-habitats.

The branching diagram that results from Camin-Sokal parsimony, Ward's Method or UPGMA are identical (Fig. 8.4). The resulting analysis must be interpreted with caution because of the large number of coefficients that were not significantly different from random. The cluster containing the sandy and the ríos Paraguay and Apa shore habitats is an example. The Río Paraguay shore habitat had similarities that were greater than can be expected at random with both sandy habitats and Río Apa shore environments (Table 8.6). The 20% similarity displayed between the latter two, however, was within random expectations. This cluster indicates that species are shared between either sandy habitat or Río Apa shore habitats and with different species found along Río Paraguay shores. For example, though the Río Paraguay shores share 11 and 12 species with Río Apa shores and sandy habitats, respectively, only five are found in all three macro-habitats. Three species (*Senna scabriuscula*, *Salix humboldtiana* var. *martiana*, and *Solanum* sp.) were collected only in these three habitats. The other two species, *Crataeva tapia* and *Polygonum punctatum*, were also found in lagoons and swamps.

Any interpretation of the remaining clusters is problematic. The cluster containing the lagoon and flooded forest habitats is based upon 20 species collected in both habitats. Nonetheless, the observed similarity, 37.04, is only 1.34 standard deviations above the random mean similarity. For the similarity to have been significant in the positive tail, a total of 25 species shared between flooded forests and the lagoons would be necessary. The backwater area serving as the outlier to the remaining macro-habitats (Fig. 8.4) is very reasonable because the backwaters had the fewest, if any, species in common with the other macro-habitats.

The overall result for plants does not present strong indication of a flood cycle. The flooded forest - lagoon cluster cannot be interpreted unambiguously. What we do see are two important effects. The first is a shore habitat plant community that exists beyond any differences between the Río Paraguay - Río Apa subregion. The plants of the sandy macro-habitats are positively associated with, and perhaps dependent upon, the Río Paraguay shore habitats. This association does not extend to the Río Apa shore samples. The

Table 8.6. Mean Simpson's Index of Similarity, S', among six macro-habitats for 186 species of plants in the Río Paraguay drainage. Abbreviations: BW–backwaters or behind flooded banks, FF–flooded forests or semilotic, LG–lagoons and swamps, SA–Río Apa shore, SN–sandy habitats, SP–Río Paraguay shore, n–number of species present, u–number of unique species, %u–percentage of unique species. Shore habitats include beaches. Coefficients shown in bold are significantly different from random (P<0.001).

	SP	SA	BW	FF	LG	SN
SP						
SA	**44.00**					
BW	11.11	12.00				
FF	**16.67**	**8.00**	**0.00**			
LG	31.48	12.00	**0.00**	37.04		
SN	**44.44**	20.00	0.00	25.93	22.22	
n	56	22	28	62	54	27
u	30	14	24	36	24	11
%u	53.57	63.64	85.71	58.06	44.44	40.74

second effect is that the backwater macro-habitats comprise a unique assemblage of aquatic plant species. No species were found in common between the backwaters and flooded forests, or lagoons and sandy macro-habitats.

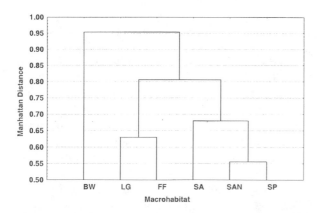

Figure 8.4. Camin-Sokal parsimony cluster analysis of macrohabiotats within the Río Paraguay based upon the planet dataset. Abbreviations: BW—backwater; FF—flooded forest; LG—lagoon; SA—shoreline Río Apa; and SP—shoreline Río Paraguay.

Discussion

The purpose of this chapter has been to compare patterns of distributions of invertebrates, plants and fishes with respect to the subregions of the Río Paraguay basin and with respect to macro-habitats. Integration of patterns among the components of biodiversity will enable us to derive the most effective conservation plan for the Río Paraguay basin between Concepción and the Brazilian border.

In an earlier chapter, Chernoff et al. (2001) were able to reject null hypotheses that the distributions of fishes were random with respect to subregions and with respect to macro-habitats. The subregional analysis demonstrated that there were two main zones: (i) a Río Paraguay zone that contained the Upper and Lower Río Paraguay subregions plus the Río Negro, and (ii) a Río Apa zone that contained the Río Apa and Riacho La Paz subregions. Within each zone species were shared broadly, between zones there was a strong faunal turnover. The macro-habitat analysis demonstrated that within the Río Paraguay zone there was a non-random association of macro-habitats due to seasonal cycles of inundation. The Río Paraguay beach habitats were central, from which most of the other interior habitats (e.g., flooded forests, backwaters, floating vegetation, and lagoons) were basically nested subsets. The deeper waters of the main channel bore the closest faunal similarity to the Río Paraguay beaches but were distant from inland habitats. Another major finding of the macro-habitat analysis was that a different faunal assemblage was present in the habitats that characterize the Río Apa zone: beaches, rapids, and clear water. This zone contains habitats more associated with terra firme and headwater areas than lowland floodplains.

The subregional analyses of macro-crustaceans and benthic invertebrates displayed almost identical results to those for fishes. The correlation of the similarity matrices was 0.92 (P<0.001). The results highlighted that there was strong faunal resemblance within the Río Paraguay zone, and that there was a rather sharp boundary to the Río Apa zone (Figs. 8.1–2). Importantly, the invertebrate result was not due to any single taxon, rather the evidence was scattered across a number of families, orders and phyla. The plant data were not presented in a way to support a full subregional analysis. We were able to test for difference between the Río Paraguay and the Río Apa zones. The aquatic plants demonstrated a strong floral boundary between the zones—only 11 out of 186 species were collected in both. Thus, our conservation recommendations, presented below, emphasize that the Río Paraguay and the Río Apa zones are highly distinctive and require separate conservation efforts.

Unlike the subregional analyses, there was less congruence among the results based upon the macro-habitats. For invertebrates, the majority of the observed similarities were significantly different from random. The pattern of similarities among macro-habitats for invertebrates was not significantly correlated with those for fishes (matrix r<0.27, P>0.05), but this lack of correlation is due to the close association in fishes between the Río Paraguay beaches and backwater habitats. Nevertheless, the clustering order of the nested sequence lagoons, flooded forests, floating vegetation, and backwaters is identical in both fishes and invertebrates (Figs. 7.11, 8.3). Furthermore, the Río Apa beaches are most different with respect to the other macro-habitats for both the invertebrates and the fishes. These results must be viewed only as a preliminary result, however, because only 23 species of invertebrates were scored for a subset of the macro-habitats for which the fishes were collected.

The plant data are difficult to interpret. Less than half of the similarity coefficients were not significantly different from random and the entropy analysis also conveyed that the matrix of species presences by macro-habitats were not significantly more ordered than a random distribution. Given these limitations, there were two aspects of the plant data that were not ambiguous. The first is that both sandy beaches and Río Apa shores share a relatively large (>10) number of species with Río Paraguay beaches. These habitats are subject to relatively stronger currents than are other habitats and may accumulate similar species. The second is that backwater samples were very different from other samples such that no species were found in common with flooded forests, lagoons, and sandy habitats.

The discovery that the invertebrate and plant data are not significantly more ordered than random expectations is due to the large number of idiosyncratic species distributions across subregions and macro-habitats. This is reflected in the relatively high number of similarity coefficients that were not significantly different from random mean similarity. The fish distributions are highly patterned and pass both tests of significance (Chernoff et al., 2001). This means that the invertebrates add weak support for the subregional analysis and for the flood-cycle relationship among macro-habitats exhibited by the fishes. This result is congruent with the results for macro-habitats exhibited by fishes and zoobenthos in the Pantanal (Chernoff and Willink, 2000). The lack of similarity between the fish and plant data is somewhat surprising given the strong association between "quiet-water" species of fishes and plants (Goulding, 1980; Lowe-McConnell, 1987; Goulding et al., 1988). The fish samples contain many species such as *Apistogramma commbrae* and *Hyphessobrycon eques* which are usually collected in association with rooted aquatic vegetation. This result may change if collections were made in closer proximity to one another.

Conclusions and Recommendations

Conservation plans must reflect departures from random distributions of the flora and fauna with respect to geography and macro-habitats. To the extent that pattern can be interpreted from the invertebrate data set, it is congruent with the non-random pattern exhibited by the fishes. The plant data provided a test that the Río Paraguay zone is different from the Río Apa zone—a finding congruent with both invertebrates and fishes. There is no confirmation by invertebrates or plants of the flood-cycle relationship among macro-habitats that was displayed by the fishes. These conclusions lead to the following recommendations:

1. The flora and the fauna comprise two major zones within the Río Paraguay basin above Concepción to the Brazilian Border: (i) the Río Paraguay zone containing the Upper and Lower Río Paraguay and the Río Negro; and (ii) the Río Apa zone containing the Río Apa and the Riacho La Paz.

2. Based upon fishes, invertebrates and plants the Río Paraguay zone contains more species than does the Río Apa zone.

3. Significant habitat within each of these zones needs to be preserved to maintain a large portion of the biodiversity.

4. There is some congruence among the fishes and invertebrates with respect to their distributions among macro-habitats but not with aquatic plants. As a result samples of all macro-habitats must be preserved to maintain the majority of species.

5. Elimination of habitats that require seasonal flooding, such as flooded forests, lagoons, and backwaters, would eliminate almost 50% of the plant species.

Literature Cited

Barbosa, F.A.R., M. Callisto, and J.A. Vianna. 2001. Water quality, phytoplankton, and benthic invertebrates of the Upper and Lower Río Paraguay basin, Paraguay. *In* Chernoff, B., P.W. Willink, and J.R. Montambault (eds.). A biological assessment of the aquatic ecosystems of the Río Paraguay basin, Departamento Alto Paraguay, Paraguay. Pp. 61–67. RAP Bulletin of Biological Assessment 19. Washington, DC: Conservation International.

Chernoff, B and P. W. Willink. 2000. Biodiversity patterns within the Pantanal, Mato Grosso do Sul, Brasil. *In* A biological assessment of the aquatic ecosystems of the Pantanal, Mato Grosso do Sul, Brasil. P. Willink, B. Chernoff, L. E. Alonso, J. R. Montambault and R. Lourival (eds.) Pp. 103–106. RAP Bulletin of Biological Assessment 18, Conservation International, Washington, DC.

Chernoff, B., P. W. Willink, M. Toledo-Piza, J. Sarmiento, M. Medina, and D. Mandelberger. 2001. Testing hypotheses of geographic and habitat partitioning of fishes in the Río Paraguay. *In* Chernoff, B., P.W. Willink, and J.R. Montambault (eds.). A biological assessment of the aquatic ecosystems of the Río Paraguay basin, Departamento Alto Paraguay, Paraguay. Pp. 80–98. RAP Bulletin of Biological Assessment 19. Washington, DC: Conservation International.

Gabriel, K. R. and R. R. Sokal. 1969. A new statistical approach to geographic variation analysis. Systematic Zoology, 18:259–270.

Goulding, M. 1980. The fishes and the forests: explorations in Amazonian natural history. Los Angeles: University of California Press. 280pp.

Goulding, M., M. L. Carvalho, and E. G. Ferreira. 1988. Rio Negro: rich life in poor water: Amazonian diversity and foodchain ecology as seen through fish communities. The Hague, Netherlands: SPB Academic Publishing. 200pp.

Lowe-McConnell, R. H. 1987. Ecological studies in tropical fish communities. Cambridge New York: University Press. 382pp.

Magalhães, C. 2001. Macro-invertebrates of the Upper and Lower Río Paraguay basin and the Río Apa sub-basin, Paraguay. *In* Chernoff, B., P.W. Willink, and J.R. Montambault (eds.). A biological assessment of the aquatic ecosystems of the Río Paraguay basin, Departamento Alto Paraguay, Paraguay. Pp. 687–71. RAP Bulletin of Biological Assessment 19. Washington, DC: Conservation International.

Mereles, M. F. 2001. Evaluation of the aquatic floral diversity in the Upper and Lower Río Paraguay basin, Paraguay. *In* Chernoff, B., P.W. Willink, and J.R. Montambault (eds.). A biological assessment of the aquatic ecosystems of the Río Paraguay basin, Departamento Alto Paraguay, Paraguay. Pp. 56–60. RAP Bulletin of Biological Assessment 19. Washington, DC: Conservation International.

Toledo-Piza, M., B. Chernoff, D. Mandelburger, M. Medina, J. Sarmiento, and P.W. Willink. 2001. Diversity and abundance of fishes in the Upper and Lower Río Paraguay basin and the Río Apa sub-basin, Paraguay. *In* Chernoff, B., P.W. Willink, and J.R. Montambault (eds.). A biological assessment of the aquatic ecosystems of the Río Paraguay basin, Departamento Alto Paraguay, Paraguay. Pp. 73–79. RAP Bulletin of Biological Assessment 19. Washington, DC: Conservation International.

Glossary

Arboreal—Pertaining to trees.

Arroyo—Stream, brook.

Basin—See *watershed*.

Benthic—Of or pertaining to the bottom of a river, lake, or other body of water.

Biodiversity—Description of the number of species, their abundance, and the degree of difference among species.

Camalote—See *floating meadows*.

Carnivorous—Organisms that feed on animals.

Cerradão—After the cerradão habitat (area consisting mainly of tree and scrub savanna with some wide, open grasslands, but mainly grassland with scattered trees and denser savanna and an occasional forest-like structure) is not exposed to fire for a while, it becomes more forest-like, thicker, and no grass.

Cerrito—Small hill.

Chaco—Also known as Gran Chaco. A generally flat region covering parts of northern Argentina, western Paraguay, and southern Bolivia and Brasil. Climate is hot and dry. Typical vegetation is grasses and cacti, with patches of thorny shrubs and deciduous dry forest.

Crepuscular—Pertaining to the morning or evening. For example, crepuscular organisms are most active in the morning and evening.

Cretaceous—A time period of the earth's history extending from about 145 million years ago to about 65 million years ago.

DBH—Diameter at breast height. Diameter of a tree at approximately 1.3 meters from the ground.

Endemic—Found only in a given area and nowhere else.

Erosion—The act of water washing away soil.

Floating meadows—Large aggregations of floating vegetation.

Fluvial—Pertaining to rivers or streams.

Herbivores—Organisms that feed on plants.

Heterogeneity—Degree of difference among items.

Hidrovia—Waterway (specifically referring to a project to straighten and channalize the Río Paraguay to give land-locked Bolivia access to the sea, and allow the passage of barges and large ocean-going vessels.

Hydrological—Pertaining to water.

Insectivores—Organisms that feed on insects.

Inundation—Flood.

Lentic—Pertaining to still water; as in lakes and ponds. See *lotic*.

Liana—Vine.

Littoral—The aquatic zone extending from the beach to the maximum depth at which light can support the growth of plants.

Lotic—Pertaining to flowing water; as in rivers and streams. See *lentic.*

Macrophyte—A non-microscopic plant.

Miocene—A time period of the earth's history extending from about 25 million years ago to about 5 million years ago.

Omnivores—Organisms that feed on a variety of food types, both plant and animal.

Pantanal—(1) A wetland. (2) Name of the world's largest wetland located primarily in southwestern Brasil, but which also occupies nearby portions of Bolivia and Paraguay.

Periphyton—Algae attached to rocks, logs, and other underwater substrates.

Physiognomy—The overall appearance or constituency of an area.

Piscivores—Organisms that feed on fishes.

Pleistocene—A time period of the earth's history extending from about 2 million years ago to about 100,000 years ago.

Potamal—The section of a river located in the lowlands and with warm temperatures.

Precambrian—A time period of the earth's history extending from the origin of the earth to about 580 million years ago.

Quaternary—A time period of the earth's history extending from about 2 million years ago to the present.

Riacho—Stream, brook.

Río—River.

Riparian—Found along the edge of a river. Often used in the context of vegetation.

Salinization—The inputing of salts into a system.

Savanna—Grasslands.

Seine—A mesh net, often used to catch fish and other larger aquatic organisms.

Semi-lotic—Pertaining to water flowing slowly. Intermediate between lentic and lotic. See *lentic* and *lotic.*

Siltation—The deposition of fine sediment (know as silt) oftentimes covering existing structures.

Stagnant—Still water (that is often foul).

Substrate—The soil found between the roots of plants or at the bottom of lakes and rivers.

Suctorial mouth—A mouth in the form of a suction-cup. It can be used to grasp logs, rocks, etc..

Watershed—A region drained by a particular river and its associated streams. Also known as a basin.

Xerophytic—Pertaining to dry, desert-like conditions.

Appendices

APPENDIX 1

Terrestrial plants recorded on transects during the AquaRAP expedition to Departamento Alto Paraguay, Paraguay, in September 1997

Hamilton Beltrán

TRANSECT 1: 19 X 10 meters.
About 1–2 kilometers before the union of the Ríos Negro and Paraguay. 20° 09.275' S, 58° 10.049' W. Riparian vegetation dominated by *Copernicia alba*. Area flooded to a depth of 30 centimeters.

FAMILY	Species	Number of Individuals > 15 cm DBH
ARECACEAE	*Copernicia alba*	20

TRANSECT 2: 30 x 10 meters.
Puerto Esperanza. 20° 18.439' S, 58° 06.220' W. Riparian vegetation dominated by *Copernicia alba*. Area flooded to a depth of 30 centimeters.

FAMILY	Species	Number of Individuals > 15 cm DBH
ARECACEAE	*Copernicia alba*	19
EUPHORBIACEAE	*Aporosella chacoensis*	1

TRANSECT 3: 60 x 10 meters.
Estancia Miranda. 20°42.210' S, 57° 59.454' W. Shrub forest.

FAMILY	Species	Number of Individuals > 15 cm DBH
ANACARDIACEAE	*Schinopsis*	1
ARECACEAE	*Copernicia alba*	7
BIGNONIACEAE	*Tabebuia nodosa*	5
CAPPARIDACEAE	*Capparis retusa*	1
FABACEAE	*Prosopis hassleri*	2
POLYGONACEAE	*Coccoloba* HB: 2687	2
SAPOTACEAE?	HB: 2693	2

TRANSECT 4: 100 x 10 meters.
Puerto Boqueron. 20° 41.591' S, 57° 56.768' W. Tall forest of 30–35 meters.

FAMILY	Species	Number of Individuals	
		> 15 cm DBH	> 30 cm DBH
BIGNONIACEAE	*Tabebuia heptaphylla*		6
FABACEAE	*Catornium polyanthum*	1	1
POLYGONACEAE	*Coccoloba* HB: 2711	7	2
RUBIACEAE	*Genipa Americana*	2	1
ULMACEAE	*Celtis* HB: 2712	1	

TRANSECT 5: 70 x 2 meters.

Puerto Lidia. 20° 46.674' S, 58° 03.947' W. 13-14 kilometers within the west bank of the Rio Paraguay. Shrub forest.

FAMILY	Species	Number of Individuals	
		> 5 cm DBH	> 15 cm DBH
ANACARDIACEAE	*Schinopsis quebracho*		1
ANACARDIACEAE	HB: 2730	1	
BIGNONIACEAE	*Tabebuia nodosa*	2	3
BOMBACACEAE	?		1
POLYGONACEAE	*Coccoloba* HB: 2727	6	
POLYGONACEAE	*Coccoloba* HB: 2728	3	
?	HB: 2733	2	
?	HB: 2734	1	

TRANSECT 6: 50 x 2 meters.

Fuerte Olimpo. 21° 02.874' S, 57° 52.710' W. Forested hills. Trees no larger than 12 meters in height. Rocky terrain.

FAMILY	Species	Number of individuals > 5 cm DBH
ANACARDIACEAE	*Astronium urundeuva*	3
BIGNONIACEAE	*Jacaranda mimosaefolia*	1
BIGNONIACEAE	*Tabebuia impetiginosa*	3
BOMBACACEAE	*Pseudobombax* HB: 2747	1
BORAGINACEAE	*Cordia glabrata*	11
RUTACEAE	HB: 27452	

TRANSECT 7: 40 x 2 meters.

Fuerte Olimpo. 21° 02.874' S, 57° 52.710' W. Forested hills. Trees no larger than 12 meters in height. Rocky terrain.

FAMILY	Species	Number of individuals > 5 cm DBH
ANACARDIACEA	*Astronium urundeuva*	3
BIGNONIACEAE	*Tabebuia impetiginosa*	11
BOMBACACEAE	*Pseudobombax* HB: 2747	3
CACTACEAE	*Stetsonia coryne*	2
MYRTACEAE	HB: 2755	1

TRANSECT 8: 80 x 2 meters.

Fuerte Olimpo. 21° 00.841' S, 57° 52.100' W. Forested hills. Trees no larger than 12 meters in height. Rocky terrain.

FAMILY	Species	Number of individuals > 5 cm DBH
APOCYNACEAE	Aspidosperma HB: 2777	11
BIGNONIACEAE	*Tabebuia* sp.	1
CACTACEAE	*Stetsonia coryne*	1
FABACEAE	HB: 2772	1
FABACEAE	HB: 2773	1
MYRTACEAE	?	2
RUTACEAE	HB: 2774	1
SAPINDACEAE	*Allophyllus* HB: 2769	1
?	HB: 2775	1

TRANSECT 9: 100 x 2 meters.

Tres Palmas. 21° 41.32' S, 57° 52.87' W. Forested hills. Trees no larger than 15 meters in height. Rocky terrain.

FAMILY	Species	Number of Individuals	
		> 5 cm DBH	> 15 cm DBH
ANACARDIACEAE	*Astronium2*	2	
FABACEAE	*Albizia*		1
FABACEAE	*Anadenanthera colubrina*	2	2
FABACEAE	?	6	
FLACOURTIACEAE	HB: 28043		
MELIACEAE	*Guarea*	1	
TILIACEAE	*Luehea divaricata*	1	

TRANSECT 10: 60 x 10 meters.

Río Apa. 10-12 kilometers above Puerto San Carlos. 22° 15.0545' S, 57° 14.097' W. Semideciduous forests of 30-35 meters in height.

FAMILY	Species	Number of Individuals	
		> 15 cm DBH	> 30 cm DBH
ANACARDIACEAE	*Astronium fraxinifolium*	3	
APOCYNACEAE	*Aspidosperma* HB: 2939		2
ARECACEAE	*Acrocomia totai*	1	
BORAGINACEAE	*Patagonula americana*	4	3
EUPHORBIACEAE	*Croton urucurana*	1	
MORACEAE	*Sorocea* HB: 2928	1	
MORACEAE	HB: 2931		2
MELIACEAE	*Trichilia* HB: 2929	1	
SAPOTACEAE	*Pouteria* HB: 2926	1	

TRANSECT 11: 57 x 2 meters.

Río Apa. 10–12 kilometers below Puerto San Carlos. 22° 13.723' S, 57° 22.638' W. Semideciduous forests of 10–15 meters in height.

FAMILY	Species	Number of Individuals	
		> 5 cm DBH	> 15 cm DBH
ANACARDIACEAE	*Astronium*		1
BORAGINACEAE	*Patagonula americana*		2
FABACEAE	*Anadenanthera colubrina*		1
MYRTACEAE	*Eugenia*	2	
SAPOTACEAE	*Pouteria* HB: 2952	1	
?	HB: 2951	13	

TRANSECT 12: 100 x 10 meters.

Riacho Blandengue. 22° 24.087' S, 57° 27.979' W. Crossing of the road of San Carlos and the Río Blandengue. Semideciduous forest of 20–25 meters in height. Terrain a gentle plain.

FAMILY	Species	Number of Individuals	
		> 15 cm DBH	> 30 cm DBH
AMARILLO	HB: 297	0	3
ANACARDIACEAE	*Astronium*	1	
BORAGINACEAE	*Patagonula americana*	1	
FABACEAE	*Anadenanthera colubrina*		1
FABACEAE	*Pterogyne nitens*		1
FASCICULOS	HB: 2972		1
RUBIACEAE	HB: 2969	5	6
SAPINDACEAE	*Allophyllus* HB: 2971	1	

APPENDIX 2

Terrestrial plants collected apart from transects during the AquaRAP expedition to Departamento Alto Paraguay, Paraguay, in September 1997

Hamilton Beltrán

Taxa
Acanthaceae
unknown sp.
Alismataceae
Echinodorus longiscapus
Amaranthaceae
Alternanthera sp.
Anacardiaceae
Astronium fraxinifolium Schott.
Schinopsis sp.
unknown sp.
Annonaceae
Rollinia emarginata? Schldl.
Apiaceae
Hydrocotyle sp.
Apocynaceae
Aspidosperma sp.
unknown sp.
Araliaceae
Pentapanax warmingianus (Marchal) Harms
Arecaceae
Bactris sp.
Copernicia alba Morong
Asclepiadaceae
unknown sp.
Asterceae
Gamochaeta sp.
Mikania sp.
Pacourina edulis Aublet
unknown sp.
Vernonia sp.
Weddelia sp.

Taxa
Bignoniaceae
Tabebuia caraiba (Mart.) Bureau
Tabebuia impetiginosa (Mart. ex DC.) Standl.
Tabebuia nodosa (Griseb) Griseb
Tabebuia sp.
unknown sp.
Bombacaceae
Pseudobombax sp.
Boraginaceae
Cordia glabrata (Mart.) A.DC.
Heliotropium sp.
Patagonula americana L.
Burseraceae
Protium sp.
Cabombaceae
Cabomba sp.
Cannaceae
Canna sp.
Capparidaceae
Capparis retusa
Capparis sp.
Capparis tuediana
Crataeva tapia L.
Caricaceae
Carica sp.
Celastraceae
Maytenus sp.
Clusiaceae
unknown sp.
Combretaceae
Combretum sp.
unknown sp.

Taxa

Commelinaceae
Commelina sp.

Convolvulaceae
Anisaea sp.
Ipomoea alba
Ipomoea sp.
Merremia? sp.
unknown sp.

Cucurbitaceae
unknown sp.
Cyperaceae
Cyperus giganteus
Eleocharis sp.
Scirpus sp.

Dilleniaceae
unknown sp.

Dioscoraceae
Dioscorea sp.

Erythroxylaceae
Erythroxylum sp.
unknown sp.

Euphorbiaceae
Acalypha sp.
Alchornea castaenifolia
Aporosella chacoensis
Cnidosculos sp.
Croton sp.
Jatropha sp.
Phyllanthus sp.
Sapium haematospermun
Sapium sp.
unknown sp.

Fabaceae
Acacia praecox Griseb
Acacia sp.
Aeschynomene sp.
Amburana cearensis (Allemao) A.C.Smith
Anadenanthera colubrina (Vell.) Brenam
Bauhinia sp.
Caesalpinia paraguariensis (D.Parodi) Burkart
Caesalpinia sp.
Cassia sp.
Cathormion polyanthum (A.Spreng.) Burkart
Enterolobium sp.
Hymenaea sp.
Inga uruguensis Hook & Arn.
Leucaena? sp.
Lonchocarpus sp.
Mimosa pellita

Taxa

Mimosa sp.
Pithecellobium sp.
Prosopis aff. Alba Griseb
Prosopis hassleri
Prosopis sp.
unknown sp.

Flacourtiaceae
Banara arguta Briq.
Casearia sp.
Prockia sp.
unknown sp.

Hippocrataceae
Salacia sp.
unknown sp.

Lamiaceae
unknown sp.

Lauraceae
Nectandra sp.
unknown sp.

Loranthaceae
Psitacanthus sp.
unknown sp.

Malphigiaceae
unknown sp.

Malvaceae
Gossypium sp.
Pavonia sp.
unknown sp.

Meliaceae
Guarea sp.
Trichilia sp.

Menispermaceae
unknown sp.

Moraceae
Chlorophora tinctoria (L.) Gaud.
Ficus enormis (Mart. ex Miq.) Miq.
Ficus sp.
Sorocea saxicola
Sorocea? sp.

Myrtaceae
Eugenia moroviana cf.
Eugenia sp.
Myrcia sp.
unknown sp.

Taxa

Nyctaginaceae
Guapira sp.
Pisonia sp.
unknown sp.

Ochanceae
Ouratea sp.

Olacaceae
Ximena sp.

Onagraceae
Caperonia? sp.
unknown sp.

Orchidaceae
unknown sp.

Oxalidaceae
Oxalis sp.

Passifloraceae
Passiflora sp.

Poaceae
Hymenachne sp.
Oryza sp.
Paspalum sp.
Schyzachirium sp.
Setaria sp.
unknown sp.

Polygalaceae
Polygala sp.

Polygonaceae
Coccoloba sp.
Polygomun sp.
Ruprechtia laxiflora Meisn
Triplaris sp.

Pontederiaceae
Pontederia cordata

Pteridophyta
Gramnitis sp.

Rhamnaceae
Gouania sp.
Ziziphus mistol Griseb

Taxa

Rubiaceae
Borreria sp.
Diodia sp.
Genipa americana L.
Palicourea sp.
Randia sp.
unknown sp.

Rutaceae
Fagara sp.
unknown sp.
Zanthoxylum sp.

Sapindaceae
Allophylus edulis
Allophylus sp.
Cardiospermun sp.
Melicocus? sp.
Paullinia sp.
unknown sp.

Sapotaceae
Bumelia obtusifolia
Pouteria sp.
Sideroxylum obtusifolium

Scrophulariaceae
unknown sp.

Simaroubaceae
Picrammia sp.

Smilacaceae
unknown sp.

Solanaceae
Physalis sp.
Solanum sp.

Sterculiaceae
Bytneria sp.
Helycteres sp.
Melochia sp.
unknown sp.
Waltheria sp.

Theoprastaceae
Clavija sp.

Tiliaceae
Luehea sp.
unknown sp.

Taxa

Ulmaceae
Celtis iguanea
Celtis sp.
Phyllostylon rhamnoides (Poiss.) Taubert
unknown sp.

Urticaceae
Urera sp.

Verbenaceae
Lantana sp.
Stachytarpeta sp.
unknown sp.
Verbena sp.

Vitaceae
Cissus sp.

Zigophyllaceae
unknown sp.

APPENDIX 3

Common names of some aquatic plants identified during the AquaRAP expedition to Departamento Alto Paraguay, Paraguay, in September 1997

María Fátima Mereles

Scientific Name	Common Name
Aporosella chacoensis	
Astronium fraxinifolium	urunde'y
Astronium urundeuva	urunde'y mí
Attalea guaranitica	palmera guasú
Azolla caroliniana	helechito de agua
Azolla filiculoides	helechito de agua
Bauhinia bauhinioides	pata de buey'í
Bromelia spp.	Caraguat*
Caesalpinia paraguariensis	guayac*n
Cathormion polyantum = *Albizia inundata*	timbó'y
Cecropia pachystachya	ambay
Celtis pubescens	yuasy'y
Chlorophora tinctoria = *Maclura tinctoria*	Mora
Crataeva tapia	yacaré pito
Croton urucurana	sangre de drago
Copernicia alba	carand*'y
Diplokeleba floribunda	palo piedra
Eichhornia azurea	aguapé, camalote
Eichhornia crassipes	aguapé puru'*, camalote
Enterolobium contortisiliquum	Timbó
Eugenia spp.	Ñangapiry
Ficus sp.	guapo'y
Genipa americana	ñandyp*
Geoffroea decorticans	Chañar
Gleditsia amorphoides	espina de corona
Guazuma ulmifolia	cambá ac*
Hymenaea coubaril	jatá yv*
Inga marginata	ingá arroyo
Juncus densiflorus	Junco
Lonchocarpus fluvialis	ka'avusú

Scientific Name	Common Name
Maytenus ilicifolia	Cangorosa
Patagonula americana	Guayaibí
Peltophorum dubium	ybyrá pyt*
Phyllostylon rhamnoides	palo lanza
Pistia stratiotes	repollito de agua
Prosopis nigra	algarrobo Negro
Prosopis ruscifolia	vinal, viñal
Prosopis vinalillo	Vinalillo
Rapanea spp.	Canelón
Salix humboldtiana	sauce criollo
Salvinia herzogii	helechito de agua
Sapium subglandulossum	curupica'y
Schinopsis balansae	quebracho colorado
Schinus fasciculatus	Molle
Stetsonia coryne	Cardón
Syagrus romanzoffiana	Pindó
Tabebuia aurea	Paratodo
Tabebuia heptaphylla	lapacho, tajy
Tabebuia nodosa	Labón
Thevetia bicornuta	jazmín del chaco
Trichilia catigua	catigu
Trichilia elegans	ca'á vovetí
Thritrinax biflabellata	Carandilla
Typha domingensis	Totora
Vitex megapotamica	Trauma
Ziziphus mistol	mistol, mbocaya'í

APPENDIX 4

Localities where aquatic plants were sampled during the AquaRAP expedition to Departamento Alto Paraguay, Paraguay, in September 1997

María Fátima Mereles

Río Paraguay Locality	Latitude	Longitude
Semi-lotic Habitats:		
1. Outlet of the Río Negro into the Río Paraguay and confluence with the Río Nabileque.	20° 09' 15.6" S	58° 10' 12.4" W
2. Puerto 14 de Mayo.	20° 18' 5.1" S	58° 06' 6.5" W
3. Puerto 14 de Mayo.	20° 17' 30.5" S	58° 06' 2.8" W
4. Puerto Boquerón.	20° 45' 51.2" S	57° 57' 29.5" W
5. Puerto Voluntad.	20° 41' 43.9" S	57° 59' 30.6" W
6. Riacho Paso Lidia.	20° 58' 34.4" S	57° 50' 19.1" W
7. Fuerte Olimpo, Riacho Vaquero.	21° 01' 03" S	57° 52' 4.0" W
8. Puerto Estancia Cerrito.	21° 26' 27.2" S	57° 54' 10.8" W
9. Estancia Cerrito, Riacho "Coe'yù".	21° 25' 12.7" S	57° 55' 21.6" W
10. Puerto 3 Palmas.	21° 52' 51.6" S	57° 57' 06" W
11. Riacho Celina.	21° 46' 34.8" S	57° 56' 56" W
12. Río Verde, in the proximity of the outlet into the Río Paraguay.	23° 06' 17.1" S	57° 42' 45.6" W
Shoreline Habitats:		
1. Outlet of the Río Negro into the Río Paraguay.	20° 09' 15.6" S	58° 10' 12.4" W
2. Estancia Miranda.	20° 40' 2.3" S	57° 59' 41.5" W
3. Estancia Cerrito, Riacho Coe'yù.	21° 25' 12.7" S	57° 55' 21.6" W
4. Estancia Cerrito.	21° 26' 27.2" S	57° 54' 10.8" W
5. Riacho Celina.	21° 46' 34" S	57° 56' 56" W
6. Puerto 3 Palmas, branch of the Río Paraguay.	21° 42' 54" S	57° 57' 24" W
7. Puerto 3 Palmas, Riacho to the south of Isla Margarita.	21° 39' 54.4" S	57° 52' 11.2" W
8. Río Napegue, vicinity of the outlet into the Río Paraguay.	22° 58' 32" S	57° 43' 38" W
9. Río Aquidabán.	23° 06' 2.8" S	57° 34' 1.8" W
Swamps:		
1. Puerto 14 de Mayo.	20° 18' 5.1" S	58° 06' 6.5" W
2. Estancia Miranda.	20° 40' 2.3" S	57° 59' 41.5" W
3. Puerto Lidia and nearby.	20° 47' 8.1" S	58° 57' 38.9" W
4. Riacho Vaquero.	20° 58' 34.4" S	57° 50' 19.1" W
5. Riacho Coè'yù.	21° 27' 46.3" S	57° 55' 40.3" W
6. Puerto 3 Palmas, Riacho Celina.	21° 46' 34.8" S	57° 56' 56.9" W
7. Puerto 3 Palmas, Riacho Celina.	21° 42' 54" S	57° 57' 24" W
Sand Banks:		
1. Banco San Alberto, outlet of Riacho San Alberto.	21° 52' 33.4" S	57° 55' 19.2" W
2. Riacho Celina.	21° 46' 34" S	57° 56' 56" W
3. Front of Estancia Cerrito.	21° 26' 27" S	57° 54' 10.8" W
Flooded Banks:		
1. Río Apa, Puerto San Carlos del Apa, 15 kilometers to the east.	22° 13' 13.9" S	57° 17' 27.2" W
2. Río Apa.	22° 13' 25.7" S	57° 14' 10.5" W
3. Río Apa.	22° 14' 0.8" S	57° 16' 17.1" W
4. Puerto San Carlos del Apa.	22° 11' 56.1" S	57° 24' 35.4" W
5. Río Apa, outlet of the Arroyo Tebicuary.	22° 12' 04" S	57° 25' 53.4" W
6. Arroyo Blandengue.	22° 24' 4.5" S	57° 27' 56" W
7. Riacho La Paz.	22° 22' 30.2" S	57° 28' 57.5" W

APPENDIX 5

Aquatic plants collected in the Río Paraguay basin, Paraguay

María Fátima Mereles

Life Style
Ep = epiphytic
Fl = free floating
La = liana
Ms = marsh
Rt = rooted semi-submergent
Sb = submergent

Taxa	Semilotic	Shoreline	Swamps	Sand Banks
Acanthaceae				
Ruellia sp.	—	Rt	—	—
Alismataceae				
Echinodorus austroamericanus	Fl	—	—	—
Echinodorus longipetalus	Rt	—	—	Rt
Echinodorus longiscapus	Rt	—	—	—
Sagittaria montevideensis	—	Rt	—	—
Amarantaceae				
Alternanthera philoxeroides	Fl	—	—	—
Alternanthera philoxeroides var. *ficoidea*	Fl	—	Fl	—
Pfaffia glomerata	Ms	—	Rt	—
Apiaceae				
Hydrocotyle ranunculoides	Fl	—	—	—
Hydrocotyle sp.	Fl	—	—	—
Apocynaceae				
Rhabdadenia ragonesei	—	—	La	—
Araceae				
Pistia stratiotes	Fl	Fl	Fl	Fl
Arecaceae				
Copernicia alba	Ms	—	—	—
Azollaceae				
Azolla caroliniana	—	—	Fl	—
Azolla filiculiodes	—	—	Fl	—

Taxa	Semilotic	Shoreline	Swamps	Sand Banks
Boraginaceae				
Heliotropium sp.	—	Rt	—	—
Butomaceae				
Hydrocleis nymphoides	—	—	Rt	—
Cabombaceae				
Cabomba australis	Sb	—	Sb	—
Cabomba sp.	Sb	—	—	—
Egeria sp.	Sb	—	—	—
Utricullaria sp.	—	—	Sb	—
Cannaceae				
Canna glauca	—	Rt	—	—
Capparaceae				
Crataeva tapia	—	Rt	Rt	Rt
Combretaceae				
Combretum lanceolatum	Ms	La	La	—
Commelinaceae				
Commelina platyphylla	—	Rt	—	—
Commelina sp.	—	—	Rt	—
Compositae				
Eclipta prostrata	Ms	—	—	Rt
Enhydra anagallis	—	Rt	—	—
Erygeron sp.	Ms	—	—	—
Mikania cordifolia	Ms	Rt	—	—
Mikania periplocifolia	Ms	La	La	La
Mikania trachypleura	—	Rt	La	—
Pacourina edulis	—	Fl	—	—
Tessaria dodonaeifolia	—	—	—	Rt
Tessaria integrifolia	—	—	—	Rt
Vernonia sp.	—	Rt	—	—
Wulffia baccata	—	—	Rt	—
Convolvulaceae				
Evolvulus sp.	—	—	La	—
Ipomoea amnicola	—	La	—	—
Ipomoea carnea spp. *fistulosa*	Ms	Rt	Rt	—
Ipomoea chiliantha	—	La	—	—
Ipomoea sp.	—	La	—	—
Ipomoea sp.	—	—	La	—
Ipomoea sp.	—	—	—	La

Taxa	Semilotic	Shoreline	Swamps	Sand Banks
Cyperaceae				
Cyperus giganteus	—	Rt	—	—
Cyperus odoratus	—	Rt	—	Rt
Cyperus surinamensis	—	—	—	Rt
Eleocharis elegans	—	—	Rt	—
Eleocharis nodulosa	—	—	Rt	—
Eleocharis sp.	—	Rt	—	—
Scirpus cubensis var. *paraguariensis*	Ep	—	—	—
Scirpus sp.	—	Rt	—	—
Euphorbiaceae				
Alchornea triplinervia	—	Rt	—	—
Aporosella chacoensis	—	Rt	Rt	—
Caperonia palustris	—	Rt	Rt	—
Phyllanthus fluitans	Fl	—	—	—
Sapium haematospermum	—	Rt	—	—
Hepatica				
Ricciocarpus natans	Fl	—	Fl	Fl
Lamiaceae				
Hyptis cfr. *mutabilis*	—	—	—	Rt
Lauraceae				
Ocotea dyospirifolia	—	—	—	Rt
Leguminosae				
Acacia caven	Ms	—	—	—
Aeschynomene falcata	Ms	—	—	—
Aeschynomene sensitiva	—	Rt	—	—
Bauhinia bauhinioides	—	Rt	Rt	—
Cathormion polyanthum	—	Rt	—	—
Dischollobium cfr. *paraguariensis*	—	Rt	—	—
Dischollobium pulchellum	Ms	—	—	—
Lonchocarpus fluvialis fluvialis	—	Rt	—	—
Macroptilium lathyroides	—	Rt	—	—
Macroptilium sp.	—	Rt	—	—
Mimosa pellita	—	Rt	Rt	Rt
Neptunia frutescens	Fl	—	—	—
Neptunia pubescens	Fl	—	Rt	—
Senna morongii	Ms	—	—	—
Senna pendula var. *paludicola*	Ms	—	—	Rt
Senna scabriuscula	—	—	—	Rt
Senna sp.	—	—	Rt	—
Sesbania virgata	—	Rt	Rt	—

Taxa	Semilotic	Shoreline	Swamps	Sand Banks
Lemnaceae				
Lemna sp.	Fl	—	—	—
Spirodela cfr. *intermedia*	Fl	—	—	—
Wolffiella sp.	Fl	—	—	—
Loranthaceae				
Psittacanthus cordatus	—	La	—	—
Malvaceae				
Hibiscus furcellatus	Ms	—	Rt	—
Hibiscus striatus	Ms	—	—	—
Hibiscus sp.	Fl	—	—	—
Hibiscus sp.	—	—	Rt	—
Marantaceae				
Thalia geniculata	—	Rt	Rt	—
Menispermaceae				
Cissampelos pareira	—	—	La	—
Menyanthaceae				
Nymphoides humboldtianum	Rt	—	Rt	—
Nymphaeaceae				
Nymphaea cfr. *gardneriana*	Rt	—	—	—
Onagraceae				
Ludwigia decurrens	—	—	—	Rt
Ludwigia helminthoriza	Fl	—	Fl	—
Ludwigia neograndiflora	—	Rt	—	Rt
Ludwigia octovalvis	—	—	—	Rt
Ludwigia peploides	Rt	—	Rt	—
Ludwigia peruviana	Rt	—	—	—
Parkeriaceae				
Salvinia herzogii	Fl	—	—	—
Salvinia natans	Fl	—	—	—
Ceratopteris paraguariensis	—	—	Fl	—
Ceratopteris cfr. *paraguariensis*	Fl	—	—	—
Passifloraceae				
Passiflora coerulea	Ms	—	—	—
Passiflora mooreana	—	—	La	—

Taxa	Semilotic	Shoreline	Swamps	Sand Banks
Poaceae				
Eragrostis hypnoides	—	Rt	—	—
Hymenachme amplexicaulis	Fl	—	Fl	—
Oryza sp.	—	—	Rt	—
Panicum elephantipes	Fl	—	—	—
Panicum sp.	—	—	Fl	—
Paspalum repens	Fl	—	Fl	—
Polygonaceae				
Coccoloba guaranitica	—	Rt	—	—
Muhelembeckia sagittifolia	—	—	—	La
Polygonum hidropiperoides	—	—	Fl	—
Polygonum punctatum	—	Rt	Rt	Rt
Polygonum stelligerum	—	Rt	Fl	—
Triplaris cfr. *guaranitica*	—	Rt	—	—
Pontederiaceae				
Eichhornia azurea	Fl	—	Fl	—
Eichhornia crassipes	Fl	Fl	Fl	—
Heteranthera limosa	Fl	—	—	—
Heteranthera multiflora	Fl	—	Fl	—
Pontederia cordata var. *cordata*	—	Rt	—	—
Pontederia rotundifolia	Fl	Fl	Fl	—
Pontederia subovata	Fl	—	—	—
Rubiaceae				
Diodia sp.	—	—	Rt	—
Genipa americana	—	Rt	—	—
Salicaceae				
Salix humboldtiana var. *martiana*	—	—	—	Rt
Salviniaceae				
Azolla caroliniana	Fl	—	—	—
Azolla filiculoides	Fl	—	—	—
Salvinia hertzogii	—	—	Fl	—
Salvinia natans	—	—	Fl	—
Sapindaceae				
Sapindus saponaria	—	Rt	—	—
Scrophulariaceae				
Angelonia sp.	—	Rt	—	—
Scoparia montevideensis	—	Rt	—	Rt
Scoparia sp.	—	Rt	—	—

Taxa	Semilotic	Shoreline	Swamps	Sand Banks
Solanaceae				
Physalis pubescens	—	—	—	Rt
Physalis sp.	—	—	—	Rt
Solanum angustifidum	—	—	Rt	—
Solanum glaucophyllum	Ms	—	—	—
Solanum spp.	Ms	—	—	—
Solanum sp.	—	—	La	—
Solanum sp.	—	—	—	Rt
Sterculiaceae				
Byttneria filipes	Ms	—	—	—
Typhaceae				
Typha domingensis	—	Rt	—	—
Urticaceae				
Urera aurantiaca	—	Rt	—	—
Utricullariaceae				
Utricullaria foliosa	Sb	—	—	—
Utricullaria sp.	Sb	—	—	—
Verbenaceae				
Vitex megapotamica	—	Rt	—	—
Vitaceae				
Cissus cicyoides var. *cicyoides*	Ms	La	—	La
Cissus verticillata	Ms	La	La	—
Total = 147 species	**62**	**56**	**54**	**27**

APPENDIX 6

Aquatic plants collected in the Río Apa sub-basin, Paraguay

María Fátima Mereles

Taxa	Flooded Banks	Along Shore
Acanthaceae		
Justicia sp.	-	■
Adiantaceae		
Adiantum orbignianum	-	■
Anacardiaceae		
Acrocomia sp.	■	-
Astronium urundeuva	■	-
Attalea sp.	■	-
Annonaceae		
Annona sp.	■	-
Apiaceae		
Hydrocotyle ranunculoides	-	■
Arecaceae		
Sheelea sp.	-	■
Bignoniaceae		
Tabebuia aurea	■	-

Taxa	Flooded Banks	Along Shore
Capparaceae		
Crataeva tapia	-	■
Cecropiaceae		
Albizzia hassleri	■	-
Cecropia pachystachya	■	-
Combretaceae		
Combretum lanceolatum	-	■
Cyperaceae		
Eleocharis cfr. *debilis*	-	■
Rhynchospora corymbosa	-	■
Euphorbiaceae		
Croton urucurana	-	■
Hygrophyllaceae		
Hygrophyla guianensis	-	■

APPENDIX 7

Localities where water quality and benthic macro-invertebrates were assessed during the AquaRAP expedition to Departamento Alto Paraguay, Paraguay, in September 1997

Francisco Antonio R. Barbosa, Marcos Callisto Faria Pereira, Pablo Moreno Souza Paula, and Juliana de Abreu Vianna

Locality	Latitude	Longitude	Date	Time
1. Río Negro (mouth), west bank.	20°09'7.9"S	58°10'11.4"W	Sep. 4	15:53
2. Río Paraguay, west bank.	20°18'05"S	58°06'6.5"W	Sep. 5	12:20
3. Río Paraguay, west bank.	20° 17'30.5"S	58°06'2.8"W	Sep. 5	—
4. Río Paraguay, (arm at Estancia Miranda), both banks.	20°40'2.3"S	57°59'41.5"W	Sep. 6	9:00
5. Río Paraguay (arm), north bank.	20°40'33.7"S	57°59'19.6"W	Sep. 6	10:30
6. Río Paraguay (Miranda), west bank.	20°41'23.2"S	57°59'40"W	Sep. 6	14:25
7. Río Paraguay, (arm, Puerto Voluntad), north bank.	20°41'33.4"S	57°56'45.6"W	Sep. 6	16:00
8. Río Paraguay (arm; cul de sac).	20°45'51.2"S	57°57'29.5"W	Sep. 7	10:00
9. Río Paraguay, (main channel, within a swamp).	20°47'8.1"S	57°57'38.9"W	Sep. 7	12:15
10. Río Paraguay, (arm; Puerto Lidia), west bank.	20°52'3.4"S	57°56'2.8"W	Sep. 8	9:15
11. Río Paraguay, (Isla Pana, main channel), east bank.	20°52'22.2"S	57°54'5.2"W	Sep. 8	10:00
12. Río Paraguay (Riacho Vaquero).	20°58'34.4"S	57°50'19.1"W	Sep. 8	15:30
13. Río Paraguay (arm in Fuerte Olimpo).	21°1'0.3"S	57°52'4.0"W	Sep. 9	8:15
14. Río Paraguay (Estancia Cerrito), east bank.	21°26'27.2"S	57°54'10"W	Sep. 10	8:15
15. Riacho Coe'yù, (west bank Río Paraguay).	21°25'12.7"S	57°55'21.6"W	Sep. 10	10:10
16. Laguna (Estancia Cerrito, littoral zone).	21°27'46.3"S	57°55'40.3"W	Sep. 10	12:15
17. Río Paraguay (beach in front of Estancia Cerrito), east bank.	21°27'26.5"S	57°55'22.8"W	Sep. 10	14:15
18. Riacho María, (arm Río Paraguay), west bank.	21°34'54.4"S	57°56'20.6"W	Sep. 11	8:50
19. Stream below Isla Margarita, east bank Río Paraguay.	21°39'54.4"S	57°52'11.2"W	Sep. 11	11:55
20. Arm of Río Paraguay, west bank.	21°42'54.7"S	57°57'24.2"W	Sep. 11	—
21. Riacho Celina (west bank Río Paraguay).	21°46'34.8"S	57°56'56.9"W	Sep. 12	8:50
22. Riacho San Alberto (mouth; beach), west bank Río Paraguay.	21°52'33.4"S	57°55'19.2"W	Sep. 12	10:50
23. Riacho San Alberto, west bank Río Paraguay (flooded palm stands).	21°52'51.6"S	57°57'0.6"W	Sep. 12	12:30
24. Stream at north bank of Río Apa (Brasil).	22°13'13.9"S	57°17'27.2"W	Sep. 14	9:30
25. Río Apa (mouth of the stream).	21°1'0.3"S	57°52'4.0"W	Sep. 14	10:20
26. Río Apa, south bank.	22°13'25.7"S	57°14'10.5"W	Sep. 14	11:30
27. Río Apa, south bank.	22°14'0.8"S	57°16'17.1"W	Sep. 14	13:30

APPENDIX 8

Benthic fauna recorded from Departmento Alto Paraguay, Paraguay during the AquaRAP expedition

Francisco Antonio R. Barbosa, Marcos Callisto Faria Pereira, Pablo Moreno Souza Paula, and Juliana de Abreu Vianna

Numbers of larvae and pupa are of *Chironomidae*.

Site	Chiro-nomidae	Larva	Pupa	Cerato-pogonidae	Chaoborus	Tipulidae	Oligochaeta	Hirudinea	Bivalvia	Gastropoda
2	83	8	—	—	—	—	16	—	—	—
3	86	—	—	—	—	—	26	—	—	—
3B	204	5	—	3	—	—	68	12	—	—
4B	2	—	—	—	—	—	13	—	—	—
5B	54	12	—	—	—	—	63	1	—	—
6B	7	—	—	2	—	—	16	—	—	—
7B	—	2	—	—	—	—	16	—	—	—
8B	6	1	—	—	—	—	50	11	1	1
10B	39	—	—	4	—	—	11	—	—	—
11B	—	—	—	—	—	—	11	—	—	—
12B	3	—	—	—	—	—	12	—	—	—
13B	—	—	—	—	—	—	137	1	—	—
14B	1	—	—	—	—	—	17	3	13	—
15B	20	—	—	—	—	—	21	—	—	—
16B	—	—	—	—	—	—	83	—	—	—
17B	9	—	1	—	—	—	—	—	—	—
18B	27	—	—	1	—	—	21	—	—	—
19B	—	—	—	—	1	—	6	—	—	—
21B	1	—	—	—	1	—	—	—	—	—
22B	4	—	—	1	—	—	10	—	—	—

Platyhel-minthes	Nematoda	Ostracoda	Concho-straca	Copepoda	Ephemer-optera	Trichoptera	Coleoptera	Odonata	Corixidae	
—	1	3	—	—	2	1	—	—	—	
—	—	—	—	—	—	—	—	—	—	
—	7	—	3	1	29	—	—	1	—	
—	—	—	—	—	—	—	—	—	—	
2	3	—	1	—	—	—	—	—	—	
—	—	—	—	—	2	—	—	—	—	
—	1	—	—	—	—	—	—	—	—	
1	—	—	—	—	—	1	—	—	—	
—	—	1	—	—	—	—	—	—	—	
—	—	—	—	—	—	—	—	—	—	
—	—	—	—	—	—	1	—	—	—	
—	—	—	—	—	—	—	—	—	—	
—	—	—	1	—	—	—	—	—	—	
—	—	—	—	—	—	—	—	—	—	
—	—	—	—	—	—	—	—	—	—	
3	—	—	1	—	7	—	—	—	—	
—	—	—	—	—	—	—	—	—	—	
—	—	—	—	—	—	—	—	—	—	
—	—	—	—	—	—	—	—	—	—	

APPENDIX 9

Sediment types, richness (R), eveness (E), and diversity (H') indices of Shannon-Wiener (base 2), total density (D (individuals/m^2)), and characteristic genera of Chironomidae

Francisco Antonio R. Barbosa, Marcos Callisto Faria Pereira, Pablo Moreno Souza Paula, and Juliana de Abreu Vianna

Sampling site	Sediment Types	R	E	H'	D	Characteristic Genera
#2 Río Paraguay	4	82	0.60	1.81	1215	*Chironomus, Nimbocera Cryptochironomus*
#3 Río Paraguay	8	79	0.00	0.00	1170	*Chironomus*
#3a Río Paraguay	9	202	0.58	1.83	2993	*Nimbocera, Polypedilum, Asheum, Ablabesmyia, Beardius*
#4 Río Paraguay	10	2	1.00	1.00	30	*
#5 Río Paraguay	9	47	0.66	2.18	696	*Nimbocera, Polypedilum*
#6 Río Paraguay	10	8	0.95	1.91	119	*
#7 Río Paraguay	1	2	0.00	0.00	30	*Chironomus*
#8 Río Paraguay	8	5	0.87	1.37	74	*
#10 Río Paraguay	9	39	0.78	2.48	578	*Nimbocera, Polypedilum*
#12 Riacho Vaquero	2	1	0.00	0.00	15	*
#14 Río Paraguay	10	1	0.00	0.00	15	*
#15 Río Caepú	9	20	0.77	1.98	296	*
#17 Laguna at Est. Cerrito	3	9	0.88	1.75	133	*
#18 Riacho María	9	25	0.80	1.85	370	*
#21 Riacho Celina	10	1	0.00	0.00	15	*
#22 Riacho San Alberto	6	5	0.96	1.52	74	*
#23 Riacho San Alberto	10	15	0.35	0.35	222	*Chironomus*
#24 Río Apa	5	38	0.65	1.30	563	*Fissimentum, Nimbocera, Coelotanypus*
#25 Río Apa	5	17	0.83	1.66	252	*Polypedilum*
#26 Río Apa	5	34	0.88	2.48	504	*Polypedilum*
#27 Río Apa	5	72	0.72	2.38	1067	*Nimbocera, Polypedilum*
#28 Río Apa	5	72	0.58	1.83	1067	*Nimbocera, Polypedilum*
#29 Río Apa	5	6	0.96	1.92	89	*
#30 Arroyo Blandengue	7	82	0.79	3.09	1215	*Goeldichironomus, Stenochironomus*
#31 Riacho La Paz	7	150	0.77	3.14	2222	*Nimbocera, Ablabesmyia, Polypedilum, Parachironomus*
#31a Riacho La Paz	7	8	0.67	1.06	119	*Polypedilum*
#35 Río Aquidabán	6	48	0.58	1.34	711	*Polypedilum*

* very low diversity of chironomids, with no representative genus.

Types of sediment: (1) silt + clay; (2) silt + filamentous algae; (3) silt + mosses; (4) silt + macrophyte's detritus; (5) silt + dry leaves; (6) fine sand; (7) coarse sand + dry leaves; (8) aquatic macrophytes + filamentous algae; (9) high content of particulate organic matter; (10) decomposing organic matter.

APPENDIX 10

Specific composition of the macro-invertebrate fauna collected during the AquaRAP expedition to the Río Paraguay and Río Apa, September 1997

Célio Magalhães

SPECIES	SUB-BASINS COLLECTED			
	Upper Río Paraguay	Lower Río Paraguay	Río Apa	Riacho La Paz
Annelida: Hirudinea				
Crustacea: Decapoda				
Palaemonidae				
Macrobrachium amazonicum	■	■	■	—
Macrobrachium borellii	—	■	■	—
Macrobrachium brasiliense	—	—	■	■
Macrobrachium jelskii	—	■	—	—
Palaemonetes ivonicus	—	■	—	—
Pseudopalaemon sp.	—	■	—	—
Sergestidae				
Acetes paraguayensis	—	■	—	—
Trichodactylidae				
Dilocarcinus pagei	■	■	—	—
Poppiana argentiniana	■	—	—	—
Sylviocarcinus australis	■	■	■	—
Trichodactylus borellianus	■	■	■	■
Valdivia camerani	■	—	—	—
Zilchiopsis oronensis	■	■	—	—
Crustacea: Conchostraca				
Cyclestheriidae				
Cyclestheria hislopi	■	■	—	—
Crustacea: Ostracoda	■	■	—	—
Crustacea: Amphipoda				
Hyallelidae				
Hyallela curvispina	—	■	—	—
Crustacea: Isopoda				
Cymothoidae				
Gen. et sp. unindentified	—	■	—	—
Crustacea: Branchiura				
Argulidae				
Dolops carvalhoi	■	■	—	—
Dolops longicauda	—	■	—	—

SPECIES	SUB-BASINS COLLECTED			
	Upper Río Paraguay	Lower Río Paraguay	Río Apa	Riacho La Paz
Mollusca: Gastropoda				
Ancylidae				
Gundlachia sp.	■	■	—	—
Ampullariidae				
Marisa planogyra	■	—	—	—
Pomacea canaliculata	■	■	—	—
Pomacea sp.	—	—	■	—
Planorbidae				
Biomphalaria sp.	■	■	■	—
Solaropsidae				
Solaropsis sp.	—	—	■	—
Succineidae				
Omalonyx sp.	■	■	—	—
Mollusca: Bivalvia				
Hyriidae				
Gen. et sp. unindentified	■	■	■	—
Anodontites sp.	—	—	■	■
Anodontites crispatus tenebricosus	—	■	■	—
Platyhelminthes: Trematoda				
Temnocephalidae	■	■	■	—
Insecta: Coleoptera				
Carabidae				
Gen. et sp. unindentified	—	■	■	—
Chrysomelidae				
Gen. et sp. unindentified	■	—	—	—
Curculionidae				
Gen. et sp. unindentified	■	■	—	—
Dryopidae				
Dryops sp.	—	—	■	—
Dytiscidae				
Gen. et sp. unindentified	■	—	—	—
Celina sp.	■	■	—	—
Copelatus sp.	■	—	—	—
Laccophilus sp. 1	■	■	■	—
Laccophilus sp. 2	—	■	—	—
Laccophilus sp. 3	—	■	—	—
Megadytes sp. 1	■	■	—	—
Megadytes sp. 2	■	—	—	—
Pachydrus sp.	■	—	—	—
Thermonectus sp.	■	■	■	—
Lampyridae				
Gen. et sp. unindentified	■	—	—	—
Noteridae				

SPECIES	SUB-BASINS COLLECTED			
	Upper Río Paraguay	Lower Río Paraguay	Río Apa	Riacho La Paz
Canthydrus sp. 1	■	—	—	—
Canthydrus sp. 2	■	—	—	—
Canthydrus sp. 3	■	—	—	—
Hydrocanthus sp. 1	■	—	■	—
Hydrocanthus sp. 2	■	—	—	—
Mesonoterus sp.	■	—	—	—
Suphis sp. 1	■	—	—	—
Suphis sp. 2	■	—	—	—
Gyrinidae				
Gen. et sp. unindentified	—	■	—	—
Gyrinus sp.	—	—	■	—
Hydrophilidae				
Gen. et sp. unindentified	■	—	■	—
Hydrobiinae	■	■	■	—
Berosus sp.	■	—	—	—
Derallus sp.	■	■	—	—
Hydrochus sp.	—	■	—	—
Tropisternus sp. 1	■	■	■	*n*
Tropisternus sp. 2	■	■	■	—
Tropisternus sp. 3	—	—	■	—
Scarabaeidae				
Dynastinae	■	—	—	—
Scirtidae				
Gen. et sp. unindentified	■	■	—	—
Insecta: Dermaptera				
Gen. et sp. unindentified	■	—	—	—
Insecta: Diptera				
Ceratopogonidae				
Gen. et sp. unindentified	—	■	—	—
Chironomidae				
Gen. et sp. unindentified	■	—	—	—
Chironominae	■	—	—	—
Orthocladinae	■	—	—	—
Tanypodinae	■	■	—	—
Culicidae				
Gen. et sp. unindentified	■	—	■	—
Insecta: Ephemeroptera				
Baetidae				
Gen. et sp. unindentified	—	—	■	—
Baetis sp.	—	—	■	■
Caenidae				
Caenis sp.	■	—	■	—
Leptophlebiidae				

SPECIES	SUB-BASINS COLLECTED			
	Upper Río Paraguay	Lower Río Paraguay	Río Apa	Riacho La Paz
Farrodes sp.	—	—	■	—
Tricorythidae				
Tricorythodes sp.	—	—	■	—
Insecta: Hemiptera				
Belostomatidae				
Belostoma bergi	■	—	■	—
Belostoma dentatum	—	—	—	■
Belostoma denticolle	■	—	—	—
Belostoma pygmeum	■	■	■	—
Belostoma triangulum	■	■	—	—
Belostoma cf. *venezuelae*	—	—	—	■
Corixidae				
Tenagobia sp.	■	—	■	—
Gelastocoridae				
Gen. et sp. unindentified	—	—	■	—
Naucoridae				
? *Pelocoris* sp.	—	■	■	—
Nepidae				
Ranatra sp.	—	—	■	—
Ranatra parvula	■	—	—	—
Ranatra cf. *parvula*	—	—	■	—
Insecta: Lepidoptera				
Family unidentified	—	■	—	—
Pyralidae				
Gen. et sp. unindentified	■	—	—	—
Insecta: Megaloptera				
Corydalidae				
Corydalus sp.	—	—	■	■
Insecta: Odonata				
Aeshnidae				
Coryphaeschna adnexa	■	■	—	—
Calopterigidae				
Hetaerina sp.	—	—	■	■
Coenagrionidae				
Gen. et sp. unindentified	■	■	—	—
? *Acanthagrion* sp.	—	—	■	—
Argia sp.	■	—	—	—
Ishnura sp.	■	—	—	—
? *Telebasis* sp.	—	■	—	—
Gomphidae				
Cyanogomphus sp.	—	—	■	—
Phyllocycla sp.	—	■	■	■

SPECIES	SUB-BASINS COLLECTED			
	Upper Río Paraguay	Lower Río Paraguay	Río Apa	Riacho La Paz
Progomphus sp.	—	—	■	■
Libellulidae				
Brachymesia sp.	■	—	—	—
Elasmothemis sp.	—	—	■	—
Erythemis sp.	■	—	—	—
? *Macrothemis* sp.	—	—	■	—
Miathyria sp.	■	■	—	—
Micrathyria sp.	■	■	—	—
Perithemis sp.	■	—	■	■
Tramea sp.	—	■	—	—
Insecta: Plecoptera				
Perlidae				
Anacroneuria sp.	—	—	■	—
Insecta: Trichoptera				
Hydropsichidae				
Gen. et sp. unindentified	—	—	■	—
Phylopotamidae				
Gen. et sp. unindentified	—	—	—	■

APPENDIX 11

Distribution of the decapod crustacean and mollusc fauna according to the habitat sampled during the AquaRAP expedition to the Río Paraguay and Río Apa, September 1997

Célio Magalhães

SPECIES	Beach Río Paraguay	Beach Río Apa	Back-waters	Clear-water	Flooded plain	Floating vegetation	Lagoon	Rapids	Cryptic
CRUSTACEA: DECAPODA									
Palaemonidae									
Macrobrachium amazonicum	■	■	■	—	■	■	■	—	—
Macrobrachium borellii	■	—	—	—	■	—	—	—	—
Macrobrachium brasiliense	—	■	—	■	■	—	—	■	■
Macrobrachium jelskii	■	—	—	—	—	—	—	—	—
Palaemonetes ivonicus						■	■	—	—
Pseudopalaemon sp.	—	■							
Sergestidae									
Acetes paraguayensis	■	—	—	—	—	—	■	—	—
Trichodactylidae									
Dilocarcinus pagei	■	—	■	—	—	■	■	—	—
Poppiana argentiniana	■	—	—	—	—	—	—	—	—
Sylviocarcinus australis	■	■	—	—	■	—	—	—	■
Trichodactylus borellianus	■	■	■	■	■	■	—	■	■
Valdivia camerani	—	—	■	—	—	■	—	—	—
Zilchiopsis oronensis	■	—	—	—	—	■	—	—	—
MOLLUSCA: GASTROPODA									
Ancylidae									
Gundlachia sp.	■	—	—	—	■	■	■	—	—
Ampullariidae									
Marisa planogyra	—	—	■	—	—	■	—	—	—
Pomacea canaliculata	■	■	■	—	■	■	■	—	—
Pomacea sp.		■	—						
Planorbidae									
Biomphalaria sp.	■	■	■	—	■	■	■	—	—
Solaropsidae									
Solaropsis sp.		■							
Succineidae									
MOLLUSCA: BIVALVIA									
Hyriidae									
Gen et sp. undetermined	—	■	■	—	■	—	■	—	—
Mycetopodidae									
Anodontites sp.	—	■	—	—	—	—	—	—	■
Anodontites crispatus tenebricosus	■	■	—	—	—	■	—	—	■

APPENDIX 12

Fish collection localities of the AquaRAP expedition to Departamento Alto Paraguay, Paraguay, in September 1997

Barry Chernoff, Darío Mandelburger, Mirta Medina, Mônica Toledo-Piza, and Jaime Sarmiento

Fishes Group A

Locality	Latitude	Longitude	Date
1. Río Negro, about 10 kilometers above mouth.	20° 5' S	58° 9' W	Sep. 4
2. Río Negro, about 1 kilometer above mouth, at east bank of river.	20° 9' S	58°10' W	Sep. 5
3. Río Negro, about 1 kilometer above mouth, at west bank of river.	20° 9' S	58° 10' W	Sep. 5
4. Río Negro, about 2 kilometers above mouth.	20° 10' S	58° 10' W	Sep. 5
5. Río Negro, Puerto Caballo, about 2 kilometers above mouth.	20° 10' S	58° 10' W	Sep. 5
6. Riacho Miranda, tributary of Río Paraguay at Puerto Miranda, about 5 kilometers from Puerto Voluntad.	20° 40' S	57° 59' W	Sep. 5
7. Riacho Miranda, about 5 kilometers from Puerto Voluntad.	20° 40' S	57° 59' W	Sep. 5
8. Río Paraguay, at west bank.	20° 17' S	58° 6' W	Sep. 5
9. Río Paraguay, at Puerto 14 de Mayo, floating vegetation at river margin.	20° 18' S	58° 6' W	Sep. 5
10. Río Paraguay, at Puerto 14 de Mayo, floating vegetation at river margin.	20° 18' S	58° 6' W	Sep. 5
11. Río Negro, at confluence with Río Paraguay.	20° 9' S	58° 10' W	Sep. 4
12. Riacho Miranda, about 1 kilometer above mouth.	20° 40' S	57° 59' W	Sep. 6
13. Riacho Miranda, about 1 kilometer above mouth, in front of Estancia Puerto Miranda.	20° 40' S	57° 59' W	Sep. 6
14. Riacho Miranda, about 1 kilometer above mouth, below Estancia Puerto Miranda.	20° 40' S	57° 59' W	Sep. 6
15. Tributary of Riacho Miranda, about 500 meters above mouth.	20° 41' S	57° 59' W	Sep. 6
16. Tributary of Riacho Miranda, about 200 meters above mouth.	20° 41' S	57° 59' W	Sep. 6
17. Riacho Miranda, above Estancia Puerto Miranda, about 1.5 kilometers above mouth.	20° 40' S	57° 59' W	Sep. 6
18. Riacho Miranda, above Estancia Puerto Miranda, about 1 kilometer above mouth.	20° 40' S	57° 59' W	Sep. 6
19. Río Paraguay, at Boqueron, backwaters near town.	20° 47' S	57° 56' W	Sep. 7
20. Riacho Lechuza, tributary of Río Paraguay below Boqueron.	20° 48' S	57° 56' W	Sep. 7
21a. Río Paraguay, upstream from Puerto Lidia.	20° 54' S	57° 55' W	Sep. 7
21b. Río Paraguay, estancia above Puerto Lidia.	20° 54' S	57° 55' W	Sep. 7
22. Río Paraguay at Puerto Lidia.	20° 54' S	57° 55' W	Sep. 7
23. Río Paraguay, above Puerto Lidia at flooded palm tree forest.	20° 53' S	57° 55' W	Sep. 8
24. Small river, above Puerto Lidia, about 200meters above mouth of small river.	20° 53' S	57° 55' W	Sep. 8
25. Río Paraguay, Estancia above Puerto Lidia.	20° 54' S	57° 55' W	Sep. 8
26. Río Paraguay, at mouth of small river in front of Fuerte Olimpo.	21o 1' S	57° 52' W	Sep. 8
27. Río Paraguay, at mouth of small river in front of Fuerte Olimpo.	21o 1' S	57° 52' W	Sep. 8
28. Río Paraguay in front of Fuerte Olimpo.	21o 2' S	57° 52' W	Sep. 8

Locality	Latitude	Longitude	Date
29. Río Paraguay, beach in front of Fuerte Olimpo.	21o 1' S	57° 52' W	Sep. 8
30. Río Paraguay at Puerto Boqueron.	20° 47' S	57° 57' W	Sep. 6
31. Tributary of Río Paraguay at Estancia Miranda.	20° 40' S	57° 59' W	Sep. 6
32. Tributary of Río Paraguay at Estancia Miranda.	20° 40' S	57° 59' W	Sep. 5
33. Tributary of Río Paraguay at Estancia Miranda.	20° 40' S	57° 59' W	Sep. 6
34. Río Paraguay, right below mouth of Río Negro.	20° 10' S	58° 8' W	Sep. 4
35. Río Paraguay at Puerto Boqueron.	20° 47' S	57° 57' W	Sep. 6
36. Río Paraguay in front of Fuerte Olimpo.	21° 2' S	57° 52' W	Sep. 9
37. Riacho Vaquero, about 2 kilometers above Fuerte Olimpo, at base of Cerro Barrero.	21° 0' S	57° 52' W	Sep. 9
38. Riacho Vaquero, about 2 kilometers above Fuerte Olimpo, at base of Cerro Barrero.	21° 0' S	57° 52' W	Sep. 9
39. Río Paraguay, between Fuerte Olimpo and El Cerrito, west bank of river, at sandy bottom beach.	21° 9' S	57° 51' W	Sep. 9
40. Río Paraguay, about 1 kilometer above Estancia Cerrito.	21° 27' S	57° 55' W	Sep. 10
41. Río Paraguay, about 1 kilometer above Estancia Cerrito.	21° 27' S	57° 55' W	Sep. 10
42. Río Paraguay at east bank, above Estancia Cerrito.	21° 26' S	57° 55' W	Sep. 10
43. Río Paraguay, above Estancia Cerrito.	21° 27' S	57° 55' W	Sep. 10
44. Riacho Coe'y˘, in palm tree flooded forest.	21° 26' S	57° 58' W	Sep. 10
45. Río Paraguay, lagoon at Estancia Cerrito.	21° 27' S	57° 55' W	Sep. 11
46. Río Paraguay, about 20 minutes above Estancia Cerrito by boat.	21° 25' S	57° 54' W	Sep. 11
47. Río Paraguay, about 1.5 kilometers above Estancia Cerrito.	21° 27' S	57° 55' W	Sep. 11
48. Small river between Puerto María Auxiliadora and Puerto Tres Palmas.	21° 42' S	57° 58' W	Sep. 11
49. Río Paraguay, above Puerto María Auxiliadora at mouth of small river.	21° 44' S	57° 57' W	Sep. 12
50. Río Paraguay, at mouth of small river in front of Valle Mi.	22° 8' S	57° 59' W	Sep. 12
51. Río Apa, south bank, above San Carlos.	22° 13' S	57° 17' W	Sep. 13
52. Río Apa, region of rapids, about 2 hours above San Carlos by boat, at beach on south bank.	22° 14' S	57° 13' W	Sep. 14
53. Río Apa, about 2 hours above San Carlos by boat, on rocks at rapids in the middle of the river.	22° 14' S	57° 13' W	Sep. 14
54. Riacho Paso Toro, tributary at south bank of Río Apa, about 1 hr above San Carlos by boat.	22° 14' S	57° 15' W	Sep. 14
55. Río Apa, below Riacho Blandengue, at north bank.	22° 12' S	57° 23' W	Sep. 15
56. Riacho Blandengue, near mouth.	22° 14' S	57° 21' W	Sep. 15
57. Riacho Blandengue, below bridge, in the property of Antebi.	22° 24' S	57° 27' W	Sep. 16
58. Riacho La Paz at headwaters.	22° 22' S	57° 28' W	Sep. 16
59. Riacho Napegue at north bank.	22° 59' S	57° 43' W	Sep. 17

Fishes Group B

Locality	Latitude	Longitude	Date
1. Río Negro, about 3 kilometers above mouth.	20° 8' S	58° 9' W	Sep. 5
2. Río Negro, about 2.5 kilometers above mouth.	20° 9' S	58° 10' W	Sep. 5
3. Riacho Miranda, about 200 meters from Estancia Miranda.	20° 40' S	57o 59' W	Sep. 6
4. Riacho Miranda, about 4 kilometers from Estancia Miranda.	20° 38' S	58° 0' W	Sep. 6
5. Río Paraguay, near Riacho Miranda.	20° 41' S	57o 59' W	Sep. 6
6. Río Paraguay, Estancia Voluntad at Puerto Voluntad.	20° 42' S	57o 57' W	Sep. 6
7. Río Paraguay, about 700 meters from Puerto Boqueron.	20° 47' S	57o 57' W	Sep. 6
8. Río Paraguay, about 2 kilometers below Puerto Boqueron.	20° 46' S	57o 57' W	Sep. 6
9. Tributary of Río Paraguay, about 2 kilometers below Puerto Boqueron.	20° 48' S	57o 56' W	Sep. 7
10. Tributary of Río Paraguay, about 2 kilometers below Puerto Boqueron.	20° 48' S	57o 56' W	Sep. 7
11. Riacho Lechuza, below Puerto Boqueron.	20°48' S	57o 56' W	Sep. 7
12. Río Paraguay, in front of Estancia Puerto Lidia at Puerto Lidia.	20° 54' S	57o 55' W	Sep. 7
13. Río Paraguay, at Puerto Lidia.	20° 54' S	57o 55' W	Sep. 7
14. Río Paraguay, about 4 kilometers above Puerto Lidia.	20° 58' S	57o 52' W	Sep. 8
15. Río Paraguay at Puerto Fuerte Olimpo.	21° 2' S	57o 52' W	Sep. 9
16. Tributary of Río Paraguay, about 10 kilometers below Puerto Fuerto Olimpo.	21° 9' S	57o 52' W	Sep. 9
17. Riacho Santa Eulalia, below Fuerte Olimpo.	21° 8' S	57o 52' W	Sep. 9
18. Río Paraguay.	21° 6' S	57o 56' W	Sep. 9
19. Río Paraguay, at lagoon at Estancia Cerrito.	21° 27' S	57o 55' W	Sep. 10
20. Río Paraguay, in front of Estancia Cerrito.	21° 27' S	57o 55' W	Sep. 10
21. Río Paraguay, at lagoon at Estancia Cerrito.	21° 27' S	57o 55' W	Sep. 10
22. Río Paraguay, at lagoon about 200 meters from large lagoon at Estancia Cerrito.	21° 27' S	57o 55' W	Sep. 10
23. Río Paraguay, above Estancia Cerrito.	21° 26' S	57o 55' W	Sep. 10
24. Río Paraguay, above Estancia Cerrito.	21° 26' S	57° 55' W	Sep. 11
25. Río Paraguay, about 1 kilometer below Estancia Cerrito.	21° 29' S	57° 56' W	Sep. 10
26. Río Paraguay, at backwaters connecting with river.	21° 35' S	57° 55' W	Sep. 11
27. Río Paraguay, at backwaters connecting with river, about 30 meters from beach at river.	21° 48' S	57° 55' W	Sep. 12
28. Riacho Celina.	21° 49' S	57° 57' W	Sep. 12
29. Tributary of Río Paraguay, about 200 meters above confluence with Río Paraguay.	21° 52' S	57° 55' W	Sep. 12
30. Río Apa, at beach about 5 kilometers above its mouth.	22° 6' S	57° 55' W	Sep. 13
31. Río Apa, at small embayment near its mouth.	22° 7' S	57° 55' W	Sep. 13
32. Río Apa, at small embayment.	22° 7' S	57° 53' W	Sep. 13
33. Riacho Mosquito, at lagoon about 1 kilometer from Estancia Ybytupora.	22° 7' S	58° 0' W	Sep. 14
34. Riacho Puerto Esperanza at swampy area.	22° 4' S	57° 59' W	Sep. 14
35. Tributary of Río Paraguay at swampy backwater area.	22° 14' S	57° 58' W	Sep. 14
36. Tributary of Río Paraguay at swampy area.	22° 14' S	57° 58' W	Sep. 14
37. Riacho Cande, in front of Puerto Casado, at backwaters.	22° 16' S	57° 58' W	Sep. 14
38. Tributary of Río Paraguay, about 500 meters below Puerto Casado.	22° 16' S	57° 55' W	Sep. 14
39. Río Apa, at sandbanks in front area of rapids.	22° 7' S	57° 51' W	Sep. 15
40. Río Apa, at swampy area in backwaters, about 500 meters from river.	22° 7' S	57° 51' W	Sep. 15
41. Río Apa, at small embayment.	22° 7' S	57° 49' W	Sep. 15
42. Río Apa, embayment surrounded by sand banks.	22° 8' S	57° 51' W	Sep. 15
43. Riacho La Paz, near mouth.	22° 24' S	57° 48' W	Sep. 16
44. Riacho La Paz, at embayment connecting with river about 80 meters from margin.	22° 24' S	57° 48' W	Sep. 16
45. Riacho La Paz, at embayment on south side.	22° 24' S	57° 46' W	Sep. 16

Locality	Latitude	Longitude	Date
46. Riacho La Paz above mouth of Riacho Primavera.	22° 24' S	57° 43' W	Sep. 16
47. Río Paraguay at Laguna Tayara, above Puerto Itacua.	22° 27' S	57° 51' W	Sep. 16
48. Río Paraguay at Puerto Itacua.	22° 29' S	57° 50' W	Sep. 16
49. Riacho Napegue at margin.	22° 58' S	57° 43' W	Sep. 17
50. Río Paraguay at ganadera Tango, kilometer 353, at west bank.	23° 7' S	57° 38' W	Sep. 17
51. Río Aquidabán, near mouth.	23° 4' S	57° 32' W	Sep. 18

APPENDIX 13

List of fishes and the subregions where they were collected during the AquaRAP expedition to Departamento Alto Paraguay, Paraguay, in September 1997

Barry Chernoff, Darío Mandelburger, Mirta Medina, Mônica Toledo-Piza, and Jaime Sarmiento

	Río Negro	Upper Río Paraguay	Lower Río Paraguay	Río Apa	Riacho La Paz
Rajiformes					
Potamotrygonidae					
Potamotrygon histrix	—	—	—	—	■
Potamotrygon motoro	—	—	■	—	—
Lepidosireniformes					
Lepidosirenidae					
Lepidosiren paradoxa	—	■	—	—	—
Characiformes					
Anostomidae					
Leporinus friderici	—	■	—	—	—
Leporinus friderici acutidens	—	—	—	■	—
Leporinus lacustris	—	■	—	—	—
Leporinus cf. obtusidens	—	—	■	—	—
Leporinus striatus	—	—	—	■	—
Schizodon cf. dissimile	—	■	■	—	—
Characidae					
Acestrorhynchus pantaneiro	—	■	■	—	—
Aphyocharax anisitsi	■	■	■	■	■
Aphyocharax dentatus	—	■	■	■	■
Aphyocharax paraguayensis	—	—	■	■	■
Astyanax fasciatus	—	—	—	■	■
Astyanax lineatus	—	—	—	■	■
Astyanax marionae	—	—	—	■	—
Astyanax paraguayensis	—	■	■	■	■
Brachychalcinus retrospina	—	—	—	■	■
Brycon sp.	■	■	—	—	—
Bryconamericus exodon	—	■	■	■	■
Bryconamericus cf. exodon	—	—	■	—	—
Bryconamericus sp. 1	—	—	—	■	■
Bryconamericus sp. 2	—	■	■	■	■
Bryconamericus sp. 3	—	■	—	■	—
Characidae sp.	—	—	—	■	—
Characidium cf. fasciatum	■	■	■	■	■

	Río Negro	Upper Río Paraguay	Lower Río Paraguay	Río Apa	Riacho La Paz
Characidium sp. 1	—	■	—	—	—
Characidium sp. 2	—	—	—	—	■
Charax leticiae	—	—	■	■	—
Charax sp.	—	■	—	—	—
Charax stenopterus	—	—	■	—	—
Cheirodon piaba	—	■	■	■	■
Cheirodon cf. *piaba*	—	■	■	■	—
Cheirodon sp. 1	—	—	—	■	—
Cheirodon sp. 2	—	■	■	—	—
Cheirodon sp. 3	—	■	—	—	—
Cheirodon cf. *stenodon*	■	—	—	—	—
Cheirodontinae sp.	—	■	—	—	—
Clupeacharax anchoveoides	—	—	—	■	—
Ctenobrycon pelegrini	—	■	■	—	—
Galeocharax gulo	—	—	■	—	—
Gymnocorymbus ternetzi	—	■	■	—	—
Hemigrammus cf. *lunatus*	—	■	■	■	■
Hemigrammus cf. *tridens*	—	—	■	—	—
Holoshesthes pequira	■	■	■	■	■
Holoshesthes cf. *pequira*	—	■	—	—	—
Hyphessobrycon eques	—	■	■	■	—
Hyphessobrycon maxillaris	—	■	—	—	—
Hyphessobrycon sp. 1	—	—	■	—	—
Hyphessobrycon sp. 2	—	—	■	—	—
Jupiaba acanthogaster	—	—	—	■	—
Metynnis mola	■	■	■	—	—
Microcharacidium sp.	—	■	■	—	—
Moenkhausia dichroura	■	■	■	■	■
Moenkhausia intermedia	—	—	■	■	■
Moenkhausia sanctaefilomenae	—	■	■	■	—
Monotocheirodon cf. sp.	■	—	—	—	—
Myleus tiete	—	■	■	—	—
Mylossoma duriventre	—	■	—	—	—
Odontostilbe paraguayensis	■	■	■	■	■
Odontostilbe sp.	—	—	—	■	—
Piabarchus cf. *analis*	—	—	■	■	■
Piabucus melanostomus	—	■	—	■	—
Piaractus mitrei	■	■	—	—	—
Poptella paraguayensis	—	—	—	■	—
Prionobrama paraguayensis	—	■	■	■	—
Pristobrycon sp.	■	■	■	—	■
Psellogrammus kennedyi	■	■	■	■	■
Pygocentrus nattereri	■	■	■	—	—
Roeboides cf. *descalvadensis*	—	—	—	—	■
Roeboides paranensis	—	■	■	■	—
Roeboides prognathus	—	—	—	■	—
Roeboides sp.	—	■	■	—	—

	Río Negro	Upper Río Paraguay	Lower Río Paraguay	Río Apa	Riacho La Paz
Salminus maxillosus	—	■	—	—	—
Serrasalmus cf. *elongatus*	—	—	■	—	—
Serrasalmus marginatus	■	■	■	—	—
Serrasalmus spilopleura	—	—	■	—	—
Tetragonopterus argenteus	—	■	■	■	—
Triportheus n. sp. *A*	—	—	—	—	■
Triportheus nematurus	—	■	■	—	—
Triportheus paranensis	—	■	■	—	—
Xenurobrycon macropus	—	—	—	■	—
Curimatidae					
Curimatella dorsalis	■	■	■	■	—
Curimatella cf. *dorsalis*	—	■	—	■	—
Potamorhina squamoralevis	—	■	■	—	—
Psectrogaster curviventris	—	■	■	—	—
Steindachnerina brevipinna	■	■	■	■	■
Steindachnerina conspersa	■	■	■	■	—
Steindachnerina insculpta	—	—	■	■	—
Cynodontidae					
Rhaphiodon vulpinus	—	■	■	—	—
Erythrinidae					
Erythrinus erythrinus	■	■	—	—	—
Hoplias malabaricus	■	■	■	■	—
Gasteropelecidae					
Thoracocharax stellatus	—	—	—	■	—
Hemiodontidae					
Hemiodus orthonops	—	■	—	—	—
Lebiasinidae					
Pyrrhulina australis	■	■	■	■	■
Parodontidae					
Apareiodon affinis	—	■	■	■	—
Prochilodontidae					
Prochilodus lineatus	—	■	—	■	—
Siluriformes					
Ageneiosidae					
Ageneiosus cf. *brevifilis*	—	—	■	—	—
Aspredinidae					
Bunocephalus australis	—	■	■	■	—
Bunocephalus doriae	—	—	—	■	—
Auchenipteridae					
Auchenipterus nuchalis	■	—	■	—	—
Parauchenipterus cf. *galeatus*	■	■	■	—	—
Tatia aulopygia	—	—	—	■	—
Callichthyidae					
Callichthys callichthys	—	—	—	■	—
Corydoras aeneus	—	—	—	■	—

	Río Negro	Upper Río Paraguay	Lower Río Paraguay	Río Apa	Riacho La Paz
Corydoras cf. *ellisae*	—	—	—	—	■
Corydoras hastatus	—	■	■	—	—
Hoplosternum littorale	■	—	—	—	—
Lepthoplosternum pectorale	■	■	■	—	—
Lepthoplosternum cf. *pectorale*	■	—	—	—	—
Megalechis thoracata	■	■	—	—	—
Megalechis cf. *thoracata*	■	—	—	—	—
Cetopsidae					
Pseudocetopsis gobioides	—	—	—	■	—
Doradidae					
Anadoras grypus	—	■	—	—	—
Doras eigenmanni	—	■	—	—	—
Doras sp.	—	■	—	—	—
Loricariidae					
Ancistrus cf. *pirarete*	—	—	■	—	—
Ancistrus cf. *piriformis*	—	—	■	■	■
Ancistrus sp.	—	■	—	—	—
Cochliodon sp. 1	—	■	—	—	—
Cochliodon sp. 2	—	—	■	■	—
Farlowella paraguayensis	—	—	—	■	—
Hypoptopoma cf. *inexspectata*	■	■	■	■	—
Hypostomus sp. 1	—	■	■	■	—
Hypostomus sp. 2	—	■	—	—	—
Hypostomus sp. 3	—	—	—	■	—
Liposarcus anisitsi	■	■	■	■	—
Loricaria sp. 1	—	■	—	—	—
Loricaria sp. 2	—	■	—	■	■
Loricariichthys cf. *platymetopon*	■	■	—	■	—
Otocinclus Maria	—	—	—	■	■
Otocinclus vestitus	—	■	■	■	—
Otocinclus vittatus	—	■	■	■	■
Pseudohemiodon laticeps	—	—	—	■	—
Rineloricaria lanceolata	—	■	—	■	■
Rineloricaria parva	—	■	■	—	—
Rineloricaria sp.	—	—	■	■	■
Pimelodidae					
Hemisorubim platyrhynchos	—	■	■	—	—
Heptapterus sp. (nov.?)	—	—	—	■	—
Imparfinis minutus	—	—	—	■	—
Microglanis cf. *parahybae*	—	—	■	■	—
Pimelodella gracilis	■	■	—	■	—
Pimelodella laticeps	—	—	—	■	—
Pimelodella mucosa	—	—	■	■	—
Pimelodella sp.	—	■	—	—	—
Pimelodus blochii	—	—	■	—	—
Pimelodus maculatus	—	—	■	—	—
Pimelodus maculatus—blochii	—	—	■	—	—

	Río Negro	Upper Río Paraguay	Lower Río Paraguay	Río Apa	Riacho La Paz
Pimelodus sp.	■	■	■	—	—
Pinirampus pirinampu	—	—	■	—	—
Rhamdia cf. *quelen*	—	■	—	■	—
Sorubim lima	—	—	■	—	—
Gymnotiformes					
Apteronotidae					
Apteronotidae sp.	—	■	■	—	—
Apteronotus albifrons	—	—	—	■	—
Gymnotidae					
Gymnotus carapo	■	■	■	—	—
Hypopomidae					
Brachyhypopomus cf. *brevirostris*	■	■	■	—	—
Brachyhypopomus sp.	■	■	—	■	—
Rhamphichthyidae					
Rhamphichthys rostratus	■	■	—	—	—
Sternopygidae					
Eigenmannia trilineata	■	■	■	—	—
Cyprinodontiformes					
Rivulidae					
Rivulus sp.	—	—	■	—	—
Beloniformes					
Belonidae					
Potamorrhaphis eigenmanni	—	■	■	■	■
Perciformes					
Cichlidae					
Apistogramma borellii	■	■	■	—	—
Apistogramma commbrae	—	■	■	■	—
Apistogramma trifasciata	—	■	■	—	—
Astronotus crassipinnis	—	■	—	—	—
Bujurquina vittata	—	■	■	■	—
Chaetobranchopsis australe	—	■	—	—	—
Cichlasoma dimerus	■	■	■	■	—
Cichlasoma sp.	—	■	—	—	—
Crenicichla lepidota	■	■	■	■	■
Crenicichla sp.	—	■	—	—	—
Crenicichla cf. *vittata*	—	■	—	—	—
Synbranchiformes					
Synbranchidae					
Synbranchus marmoratus	■	■	■	■	—
Total = 173 species	41	105	93	85	35

APPENDIX 14

Geographic distribution of fish species collected during the AquaRAP expedition to Departmento Alto Paraguay, Paraguay

Mônica Toledo-Piza

Species identified as sp. 1, 2, etc. in Appendix 13 are not included in the present table.

An asterisk (*) before a species name indicates that it has not been previously recorded for the Río Paraguay. However, identifications should be carefully checked before identifying the species as a new record.

Two asterisks (**) before a species name indicates that it is restricted to the Río Paraguay basin.

Many species indicated as 'widespread' belong to poorly studied taxa and probably constitute species complexes. Revisionary studies may indicate that the latter includes species restricted to the Paraguay basin.

Species	Range
Rajiformes	
POTAMOTRIGONIDAE	
Potamotrygon histrix	Lower Río Paraná, Río Paraguay (Rosa, 1985).
Potamotrygon motoro	Widespread (Rosa, 1985).
Lepidosireniformes	
LEPIDOSIRENIDAE	
Lepidosiren paradoxa	Rio da Prata basin, Río Paraguay (Pantanal region), Amazon basin (Planquette et al., 1996; Britski et al., 1999).
Characifzormes	
ANOSTOMIDAE	
Leporinus friderici	Amazon, Paraguay-Paraná basins, rivers of Surinam (Garavello et al., 1992).
Leporinus lacustris	Río Paraguay, Río Paraná basins (Garavello, 1979).
Leporinus cf. *obtusidens*	Rio São Francisco, Río Paraná, Río Paraguay (Pantanal included), Río Uruguay basins (Garavello, 1979).
Leporinus striatus	Río Paraguay (Pantanal included), Río Paran, Río Magadalena, Upper Amazon basins (Britski and Garavello, 1980).
Schizodon cf. *dissimile*	Type locality: Río Poti (Río Paranaíba drainage, Piauí State).

Species	Range
Characidae	
Acestrorhynchus pantaneiro	Río Paraguay (Pantanal included), Rio da Prata, Río Mamoré basins (Menezes, 1992).
Aphyocharax anisitsi	Río Paraguay (Pantanal included), Río Guaporé (Pearson, 1937; Britski et al., 1999), Río Uruguay basins (Eigenmann, 1915).
Aphyocharax dentatus	Río Paraguay (Pantanal included), Río Guaporé basins (Pearson, 1937; Britski et al., 1999).
**Aphyocharax paraguayensis*	Río Paraguay (Pantanal included) (Pearson, 1937; Britski et al., 1999).
Astyanax fasciatus	Widespread.
Astyanax lineatus	Río Paraguay (Pantanal included), Río Parana·, Bolivia (Fowler, 1948; Britski et al., 1999).
**Astyanax marionae*	Río Paraguay (Pantanal included) (Fowler, 1948; Britski et al., 1999).
Astyanax paraguayensis	Río Paraguay and Upper Tocantins (Eigenmann, 1921).
**Bryconamericus exodon*	Río Paraguay (Pantanal included) (Eigenmann, 1927; Britski et al., 1999).
**Brachychalcinus retrospina*	Río Paraguay basin (Pantanal included) (Reis, 1989).
Characidium cf. *fasciatum*	Rio das Velhas drainage, Río Paraná basin (Buckup, 1992).
Charax leticiae	Rio Tocantins, Río Paraguay (in Pantanal region) (Lucena, 1989).
Charax stenopterus	Rio da Prata system, coastal lagoon system in Rio Grande do Sul State, lower Río Paraná (Lucena, 1987).
**Cheirodon piaba*	Distribution restricted to Rio São Francisco basin and rivers in northeastern Brasil (Malabarba, 1994).
Cheirodon cf. *stenodon*	Restricted to Upper Río Paraná. Uj's (1987) citation for the Río Paraguay is considered a misidentification (Malabarba, 1994).
Clupeacharax anchoveoides	Río Paraguay (in Pantanal region) (Britski et al., 1999), records from Madre de Dios, Rio Purus basin (NEODAT, 1998), Río Beni-Mamoré (Pearson, 1937).
**Ctenobrycon pellegrini*	Río Paraguay (Pantanal included) (Eigenmann, 1921; Britski et al., 1999) (? include in Astyanax).
**Galeocharax gulo*	Amazon and Rio São Francisco basins (the species recorded for the Paraguay is G. humeralis) (Menezes, 1976).
Gymnocorymbus ternetzi	Río Paraguay (Pantanal included) (Britski et al., 1999), Río Guaporé (Benine, 2000).
Hemigrammus cf. *lunatus*	Type locality: Amazon basin, Río Paraguay (in Pantanal region) (Eigenmann, 1918; Britski et al., 1999).
**Hemigrammus* cf. *tridens*	Río Paraguay (Pantanal included) (Eigenmann, 1918; Britski et al., 1999).
Holoshesthes pequira	Type locality: Villa Bella; Río Paraguay (Pantanal included); ? Río Paraná basins (Uj, 1987).
Hyphessobrycon eques	Río Paraguay (Pantanal included), Amazon basins (Weitzman and Palmer, 1997).
**Hyphessobrycon maxillaris*	Río Paraguay (Pantanal included) (Fowler, 1948; Britski et al., 1999) (*Hemigrammus* according to Weitzman (1985)).
Jupiaba acanthogaster	Río Paraguay (Pantanal included), Rio Tapajûs, Rio Tocantins basins (Zanata, 1997).
**Metynnis mola*	Río Paraguay (Pantanal included (Fowler, 1950; Britski et al., 1999).
* *Microcharacidium* sp.	Genus reported from Amazon and Orinoco basins, and rivers of Surinam and French Guiana. One species, Microcharacidium weitzmani, has one record in the Río Guaporé basin (Buckup, 1993).
Moenkhausia dichroura	Río Paraguay (Pantanal included), Amazon, Guyana (Eigenmann, 1917; Britski et al., 1999), Río Guaporé basins (Fowler, 1948).
Moenkhausia intermedia	Río Paraguay (Pantanal included), Amazon, Paraná basins (Eigenmann, 1917).
Moenkhausia sanctaefilomenae	Type locality: Río Paraguay (Pantanal included), Rio Parnaíba basins.
Myleus tiete	Type locality: Piracicaba, São Paulo State; Río Paraguay (Géry, 1977).
Odontostilbe paraguayensis	Río Paraguay (Pantanal included), and lower Río Parana basin.
**Piabarchus* cf. *analis*	Upper Río Paraguay (Pantanal region) (Mahnert and Géry, 1988; Britski et al., 1999).
**Piabucus melanostomus*	Río Paraguay (Pantanal included) (Vari, 1977; Britski et al., 1999).

Species	Range
Piaractus mitrei	Río Paraguay, Rio da Prata (Britski et al. (1999) consider it a junior synonym of P. mesopotamicus).
Poptella paraguayensis	Throughout Río Paraguay basin and lower Río Paraná.
Prionobrama paraguayensis	Río Paraguay (Pantanal included), Río Uruguay basins (Eigenmann, 1915; Britski et al., 1999).
Psellogrammus kennedyi	Río Paraguay (Pantanal included (Britski et al., 1999), Rio São Francisco (Fowler, 1948; Géry, 1977).
Pygocentrus nattereri	Widespread (Fink, 1993).
Roeboides cf. *descalvadensis*	Río Paraguay (Pantanal included), lower Río Paraná basins (Lucena, 1988).
Roeboides paranenesis	Río Paraguay (Pantanal included), lower and Middle Río Paraná basins (Lucena, 1988).
Roeboides prognathus	Rio da Prata system (Pantanal included) (Lucena, 1988).
Salminus maxillosus	Río Paraguay (Pantanal included), La Plata basin, Amazon basins (Fowler, 1950; Géry, 1977; Britski et al., 1999).
* Serrasalmus cf. *elongatus*	Type locality: Río Guaporé; Amazon basins (Fowler, 1950).
Serrasalmus marginatus	Type locality: Río Paraná; Río Paraguay (Pantanal included) (Britski et al., 1999).
Serrasalmus spilopleura	Type locality: Río Guaporé; probably Rio Tocantins (Jégu and Santos, 1988).
Tetragonopetrus argenteus	Throughout South American rivers (Eigenmann, 1917).
**Triportheus nematurus	Upper Río Paraguay (Pantanal region) (Portugal, 1990).
Triportheus paranenesis	Río Paraguay (Pantanal included), lower Río Paraná basin (Portugal, 1990).
Xenurobrycon macrops	Río Paraguay (Pantanal included), lower Río Paraná basin (Weitzman and Fink, 1985).
CURIMATIDAE	
Curimatella dorsalis	Río Paraguay (Pantanal included), lower Río Paraná, Amazon, Rio Tocantins, Río Orinoco basins (Vari, 1992).
Potamorhina squamoralevis	Río Paraguay, Río Paraná basins (Vari, 1984).
Psectrogaster curviventris	Río Paraguay, Río Guaporé basins (Vari, 1989).
Steindachnerina brevipinna	Río Paraguay (Pantanal included), lower Paraná, lower Río Uruguay (Vari, 1991).
Steindachnerina conspersa	Río Paraguay, lower Río Paraná (Vari, 1991).
Steindachnerina insculpta	Upper Río Paraná basin. Occurrence in Upper Río Paraguay basin is hypothesized as a recent introduction (Vari, 1991).
CYNODONTIDAE	
Rhaphiodon vulpinus	Río Amazonas, Río Orinoco, rivers of the Guianas, Río Paraná,Río Guaporé basins (Toledo-Piza, 1997).
ERYTHRINIDAE	
Erythrinus erythrinus	Type locality: Suriname; widespread.
Hoplias malabaricus	Widespread.
GASTEROPELECIDAE	
Thoracocharax stellatus	Amazon and Río de La Plata systems (Weitzman, 1954, 1960) (includes records from the Río Guaporé basin).
HEMIODONTIDAE	
Hemiodous orthonops	Río Paraguay, lower Río Paraná basins (Langeani-Neto, 1996).
LEBIASINIDAE	
Pyrrhulina australis	Río Paraguay (Pantanal included) (Britski et al., 1999), Upper Río Paraná basins.
PARODONTIDAE	
Apareiodon affinis	Río Paraguay (Pantanal included), Río Paraná, Río Uruguay basins (Pavanelli, 1999; Britski et al., 1999).
PROCHILODONTIDAE	
Prochilodus lineatus	Río Paraguay (Pantanal included), Río Paraná, Río Uruguay basins (Castro, 1990).

Species	Range
Siluriformes	
AGENEIOSIDAE	
Ageneiosus cf. *brevifilis*	Type locality: French Guiana; Amazon, Río Paraguay, Río Paraná, Río Orinoco basins (Britski, 1972).
ASPREDINIDAE	
***Bunocephalus australis*	Type locality: Corumb·; Pantanal region (Britski et al., 1999), Río Paraguay basin (Mees, 1989).
Bunocephalus doriae	Río Paraguay (Pantanal included) (Britski et al., 1999), Río Paraná, Río Uruguay, Rio Jacuí (Mees, 1989).
AUCHENIPTERIDAE	
Auchenipterus nuchalis	Widespread (Mees, 1974).
Parauchenipterus cf. *galeatus*	Widespread (Mees, 1974).
Tatia aulopygia	Río Paraguay (Pantanal included), Río Guaporé, Río Paraná basins (Soares-Porto, 1996).
CALLICHTHYIDAE	
Callichthys callichthys	Widespread.
Corydoras aeneus	Type locality: Trinidad; Río Napo, Río Ucayali basins (Nijssen and Isbrücker, 1979, 1986), Pantanal (Britski et al., 1999).
**Corydoras cf. *ellisae*	Río Paraguay (Nijssen and Isbrücker, 1979).
Corydoras hastatus	Type locality (Villa Bella = Parintins - Río Amazonas basin), Pantanal (Britski et al., 1999).
Hoplosternum littorale	Widespread throughout South American rivers including Río Paraguay and Río Guapore basins. Occurrence in Río Paraná is hypothesized as recent introduction (Reis, 1997).
***Lepthoplosternum pectorale*	Río Paraguay (Pantanal included) (Reis, 1997).
* *Megalechis thoracata*	Not recorded for the Paraguay (*Megalechis personata* - 1 locality in Upper Paraguay).
CETOPSIDAE	
Pseudocetopsis gobioides	Río Paraguay (Pantanal included), Río Paraná (Oliveira, 1988; Britski et al., 1999).
DORADIDAE	
Anadoras grypus	Type locality: Rio Ambyiacu; *A. weddelli* is recorded for the Río Paraguay (including Pantanal) basin (Higuchi, 1992; Britski et al., 1999).
Doras eigenmanni	Río Paraguay (Pantanal included) (Britski et al., 1999); ? Amazon, Bolivia (Fowler, 1951).
LORICARIIDAE	
**Ancistrus cf. *pirareta*	Type locality: Río Tebicuary, Río Paraguay basins.(Müller, 1989).
Ancistrus cf. *piriformis*	Type locality: Río Acaray, lower Río Paraná (Müller, 1989).
***Farlowella paraguayensis*	Appears to be endemic to the Río Paraguay basin (including the Pantanal region) (Retzer and Page, 1996).
**Hypoptopoma cf. *inexspectata*	Type locality: Río Paraguay (Formosa) (Isbrücker, 1980).
***Liposarcus anisitsi*	Río Paraguay (Pantanal uncluded) (Fowler, 1954; Britski et al., 1999).
Loricariichthys cf. *platymetopon*	Middle and lower Río Paraguay, lower Paraná, Río Uruguay, ? Amazon (Isbrucker and Nijssen, 1979).
Otocinclus Maríae	Río Paraguay, Río Guaporé/Mamoré, lower Amazon basins (Schaefer, 1997).
Otocinclus vestitus	Río Mamoré/Guaporé, lower Río Paraná basins (Schaefer, 1997).
Otocinclus vittatus	Widespread throughout lowland Upper Amazon, Orinoco, Paraná, and Paraguay basins; scattered collections from Ríos Xingu and Tocantins (Schaefer, 1997).
Pseudohemiodon laticeps	Lectotype from Paraguay; lower Paraná, Río Pastaza drainage (Ecuador), (Isbrücker and Nijssen, 1978).
Rineloricaria lanceolata	Amazon, Río Guaporé basins(Isbr cker, 1973).
Rineloricaria parva	Río Paraguay (Pantanal included) (Fowler, 1954; Britski et al., 1999).

Species	Range
PIMELODIDAE	
Hemisorubim platyrhynchus	Río Orinoco to La Plata basin (Mees, 1974).
Imparfinis minutus	Suriname, Río Branco drainage, Rio das Velhas drainage (Mees, 1974).
Microglanis cf. *parahybae*	Type locality: Rio Paraiba do Sul; Río Paraguay (1 record), (Mees (1974) in Britski et al. (1999)) (*Microglanis cottoides*, considered a synonym of *M. parahybae* by Mees (1974)).
Pimelodella gracilis	Type locality: Corriente, Río Paraná, Río Uruguay-Paraná, La Plata basins, records from other basin may constitute different species (Mees, 1974).
***Pimelodella laticeps*	Río Paraguay (not in Britski et al. (1999)) (Fowler, 1951).
Pimelodella mucosa	Río Paraguay, (Pantanal included), Río Mamoré (Lauzanne and Loubens, 1985).
Pimelodus blochii	Widespread.
Pimelodus maculatus	Widespread.
Pinirampus pirinampu	Widespread.
Rhamdia cf. *quelen*	Widespread (Silfvergrip, 1996).
Sorubim lima	Widespread.
Gymnotiformes	
APTERONOTIDAE	
Apteronotus albifrons	Río Paraguay, Río Paraná, Amazon basins, rivers of Guyana (Campos da Paz, 1997).
GYMNOTIDAE	
Gymnotus carapo	Specimens from Río Paraguay and Río Paraná may constitute a different species (Campos da Paz, 1997).
HYPOPOMIDAE	
Brachypopomus cf. *brevirostris*	Upper Río Paraguay (Pantanal region), ? Río Paraná, Amazon basins (Campos da Paz, 1997).
RHAMPHICHTHYIDAE	
Rhamphichthys rostratus	*Rhamphichthys hahni* is the only *Rhamphichthys* species that occurs in the Río Paraguay and Paraná basins. Presence of *R. rostratus* was not confirmed (Campos da Paz, 1997).
STERNOPYGIDAE	
Eigenmannia trilineata	Río Paraguay (Pantanal included), Río Paraná (Campos da Paz, 1997).
Beloniformes	
BELONIDAE	
Potamorrhaphis eigenmanni	Río Paraguay (Pantanal included), and Río Guaporé (Collette, 1982).
Perciformes	
CICHLIDAE	
Apistogramma borellii	Río Paraguay (Pantanal included) and lower Río Paraná basins (Kullander, 1982b).
***Apistogramma commbrae*	Río Paraguay (Pantanal included) (Kullander, 1982a).
Apistogramma trifasciata	Río Paraguay (Pantanal included), Río Guaporé, ? lower Río Paraná, (Kullander, 1982b).
Astronotus crassipinnis	Río Paraguay, Río Guaporé basins (Kullander, 1986).
***Chaetobranchopsis australe*	Río Paraguay (did not find additional information).
Crenicichla lepidota	Río Paraguay, lower Río Paraná, Middle Río Uruguay, Río Guaporé basins (Lucena and Kullander, 1992).
Crenicichla vittata	Río Paraguay, Río Uruguay, Río Paraná basins (Lucena and Kullander, 1992).
Cichlasoma dimerus	Río Paraguay (Pantanal included), lower Río Paraná (Kullander, 1983).

Species	Range
Synbranchiformes	
SYNBRANCHIDAE	
Synbranchus marmoratus	Widespread.

Literature Cited

Benine, R. 2000. Taxonomia e relaçõies filogenéticas de Gymnocorymbus Eigenmann, 1908 (Characiformes, Characidae). Unpublished Masters dissertation. Botucatu, Brasil: Instituto de Biociências—Universidade Estadual de São Paulo (UNESP). 55pp.

Britski, H.A. 1972. Sistematica e evolucao dos Auchenipteridae e Ageneiosidae (Teleostei, Siluriformes). Unpublished Masters thesis. São Paulo: Instituto Biociências da Universidade de São Paulo. 142pp.

Britski, H.A. and J.C. Garavello. 1980. Sobre uma nova espécie de *Leporinus* da bacia amazônica (Pisces, Anostomidae) com considerações sobre *L. striatus* Kner, 1850 e espécics afins. Papéis Avulsos de Zoologia, São Paulo, 35:253–262.

Britski, H. A., K. Z. S. de Silimon, and B. S. Lopes. 1999. Peixes do Pantanal: manual de identificação. Brasília, Brasil: Embrapa. 184pp.

Buckup, P.A. 1992. Redescription of *Characidium fasciatum,* type species of the Characidiinae (Teleostei, Characiformes) Copeia, 1992:1066–1073.

Buckup, P.A. 1993. Review of the characidiin fishes (Teleostei: Characiformes), with desriptions of four new genera and ten new species. Ichthyological Exploration of Freshwaters, 4:97–154.

Campos da Paz, R. 1997. Sistemática e taxonomia dos peixes elétricos das bacias dos rios Paraguai, Paraná e São Francisco, comáotas sobre espécies presentes em rios costeiros do leste do Brasil (Teleostei: Ostariophysi: Gymnotiformes). Unpublished Ph.D. dissertation. São Paulo: Universidade de São Paulo.

Castro, R.M.C. 1990. Revisão taxonômica da família Prochilodontidae (Ostariophysi: Characiformes). Unpublished Ph.D. dissertation. São Paulo: Universidade de São Paulo. 293pp.

Collette, B.B. 1982. South American freshwater needlefishes of the genus *Potamorrhaphis* (Beloniformes: Belonidae). Proceedings of the Biological Society of Washington, 95:714–747.

Eigenmann, C.H. 1915. The Cheirodontinae, a subfamily of minute characid fishes of South America. Memoirs of the Carnegie Museum 7:1–99.

Eigenmann, C.H. 1917. The American Characidae. part 1. Memoirs of the Museum of Comparative Zoology, 43:1–102.

Eigenmann, C.H. 1918. The American Characidae. part 2. Memoirs of the Museum of Comparative Zoology, 43:103–208.

Eigenmann, C.H. 1921. The American Characidae. part 3. Memoirs of the Museum of Comparative Zoology, 43:209–310.

Eigenmann, C.H. 1927. The American Characidae. part 4. Memoirs of the Museum of Comparative Zoology, 43:311–428.

Fink, W.L. 1993. Revision of the piranha genus *Pygocentrus* (Teleostei, Characiformes) Copeia, 1993:665–687.

Fowler, H.W. 1948. Os peixes de ·gua doce do Brasil (1a entrega). Arquivos de Zoologia do Estado de São Paulo, 6:1–204.

Fowler, H.W. 1950. Os peixes de gua doce do Brasil (2a entrega). Arquivos de Zoologia do Estado de São Paulo, 6:205–404.

Fowler, H.W. 1951. Os peixes de gua doce do Brasil (3a entrega). Arquivos de Zoologia do Estado de São Paulo, 6:405–625.

Fowler, H.W. 1954. Os peixes de gua doce do Brasil (4a entrega). Arquivos de Zoologia do Estado de São Paulo, 9:1–400.

Garavello, J.C. 1979. Revisão taxonômica do genero *Leporinus* Spix, 1829 (Ostariophysi, Anostomidae). Unpublished Ph.D. dissertation. São Paulo: Instituto de Biociências da Universidade de São Paulo. 451pp.

Garavello, J.C., S.F. dos Reis, and R.E. Strauss. 1992. Geographical variation in *Leporinus friderici* (Bloch) (Pisces: Ostariophysi: Anostomidae) from the Paraná-Paraguay and Amazon River basins. Zoologica Scripta, 21:197–200.

Géry, J. 1977. Characoids of the world. Neptune City, New Jersey: T.F.H. Publications, Inc. Ltd. 672pp.

Higuchi, H. 1992. A phylogeny of the South American thorny catfishes (Osteichthyes: Siluriformes, Doradidae). Unpublished Ph.D. dissertation. Cambridge, Massachusetts: Harvard University. 372pp.

Isbrücker, I.J.H. 1973. Status of the primary homonymous South American catfish *Loricaria cirrhosa* Perugia, 1897, with remarks on some other loricariids (Pisces, Siluriformes, Loricariidae). Annali Museo Civico di Storia Naturale di Genova, 79:172–191.

Isbrücker, I.J.H. 1980. Classification and catalogue of the mailed Loricariidae (Pisces, Siluriformes). Verslagen en Technische Gegevens, 22:1–181.

Isbrücker, I.J.H. and H. Nijssen. 1978. Two new species and a new genus of neotropical mailed catfishes of the subfamily Loricariinae Swainson, 1838 (Pisces, Silurifromes, Loricariidae) Beaufortia, 27:177–206.

Isbrücker, I.J.H. and H. Nijssen. 1979. Three new South American mailed catfishes of the genera *Rineloricaria* and *Loricariichthys* (Pisces, Siluriformes, Loricariidae). Bijdragen tot de Dierkunde, 48:191–211.

Jégu, M. and G.M. Santos. 1988. Le genre *Serrasalmus* (Pisces, Serrasalmidae) dans le bas Tocantins (Brésil, Par·), avec la description d'une espêce nouvelle, *S. geryi,* du bassin Araguaia-Tocantins. Revista Hydrobiologia Tropical, 21:239–274.

Kullander, S.O. 1982a. Cichlid fishes from the La Plata basin. part II. *Apistogramma commbrae* (Regan, 1906). Revue Suisse de Zoologie, 89:33–48.

Kullander, S.O. 1982b. Cichlid fishes from the La Plata basin. part IV. review of the *Apistogramma* species, with description of a new species (Teleostei, Cichlidae). Zoologica Scripta, 11:307–313.

Kullander, S.O. 1983. A revision of the South American cichlid genus *Cichlasoma* (Teleostei: Cichlidae). Stockholm: Naturhistoriska Riksmuseet. 296pp.

Kullander, S.O. 1986. Cichlid fishes of the Amazon River drainage of Peru. Stockholm: Swedish Museum of Natural History. 431pp.

Langeani-Neto, F. 1996. Estudo filogenético e revisão taxonômica da família Hemiodontidae Boulenger, 1904 (sensu Roberts, 1974) (Ostariophysi, Characiformes). Ph.D. dissertation. São Paulo: Universidade de São Paulo. 171pp.

Lauzanne, L. and G. Loubens. 1985. Peces del río Mamoré. Paris: ORSTOM. 116pp.

Lucena, C.A.S. de. 1987. Revisão e redefinição do gênero neotropical *Charax* Scopoli, 1777 com a descrição de quatro espécies novas (Pisces; Characiformes; Characidae). Comunicações do Museu de Ciências da PUCRS, 40:5–124.

Lucena, C.A.S. de. 1988. Lista comentada das espéces do gênero *Roeboides* Günther, 1864 descritas para as bacias dos rios Amazonas, São Francisco e de Prata (Characiformes, Characidae, Characinae). Comunicações do Museu de Ciências da PUCRS, Série Zoologia, Porto Alegre 1:29–47.

Lucena, C.A.S. de. 1989. Trois nouvelles espéces du genre *Charax* Scopoli, 1777 pour la région Nord du Brésil (Characiformes, Characidae, Characinae). Revue Francaise d'Aquariologie, 15:97–104.

Lucena, C.A.S. de and S.O. Kullander. 1992. The *Crenicichla* (Teleostei: Cichlidae) species of the Uruguai River drainage in Brazil. Ichthyological Exploration of Freshwaters, 3:97–160.

Mahnert, V. and J. Géry. 1988. Les genres Piabarchus Myers et *Creagrutus* Günther du Paraguay, avex la description de deux nouvelles espéces (Pisces, Ostariophysi, Characidae). Revue Francaise d'Aquariologie, 15:1–8.

Malabarba, L.R. 1994. Sistem"tica e filogenia de Cheirodontinae (Ostariophysi: Characiformes: Characidae). Unpublished Ph.D. dissertation. São Paulo: Universidade de São Paulo. 287pp.

Mees, G.F. 1974. The Auchenipteridae and Pimelodidae of Suriname (Pisces, Nematognathi). Zoologische Verhandelingen, 132:1–256.

Mees, G.F. 1989. Notes on the genus *Dysichthys,* subfamily Bunocephalinae, family Aspredinidae (Pisces, Nematognathi). Proceedings of the Koninklijke Nederlandse Akademie van Wetenschappen (Series C), 92:189–250.

Menezes, N.A. 1976. On the Cynopotaminae, a new subfamily of Charcidae (Osteichthyes, Ostariophysi, Characoidei). Arquivos de Zoologia, São Paulo, 28:1–91.

Menezes, N.A. 1992. Redefinição taxonómica das espécies de *Acestrorhynchus* do grupo *lacustris* com a descrição de uma nova espécie (Osteichthyes, Characiformes, Characidae). Comunicações do Museu de Ciências da PUCRS, Série Zoologia, Porto Alegre, 5:39–54.

Müller, S. 1989. Description de deuz nouvelles espêces paraguayennes du genre Ancistrus Kner, 1854 (Pisces, Siluriformes, Loricariidae). Revue Suisse de Zoologie, 96:885–904.

NEODAT. 1998. Inter-Institutional Database of Fish Biodiversity in the Neotropics. http://biodiversity.bio.uno.edu/~neodat/.

Nijssen, H. and I.J.H. Isbrücker. 1979. Chronological enumerations of nominal species and subspecies of *Corydoras* (Pisces, Siluriformes, Callichthyidae). Bulletin Zoologisch Museum, Universiteit van Amsterdam, 6:129–135.

Nijssen, H. and I.J.H. Isbrücker. 1986. Review of the genus Corydoras from Peru and Ecuador (Pisces, Siluriformes, Callichthyidae). Studies on Neotropical Fauna and Environment, 21:1–68.

Oliveira, J.C. de. 1988. Osteologia e revisão sistemática de Cetopsidae (Teleostei, Siluriformes). Unpublishedd Ph.D. dissertation. São Paulo: Universidade de São Paulo. 242pp.

Pavanelli, C. S. 1999. Revisão taxonômica da família Parodontidae (Ostariophysi: Characiformes). Ph.D. dissertation, Universidade Federal de São Carlos, São Carlos, Brasil. 334pp.

Pearson, N.E. 1937. The fishes of the Beni-Mamoré and Paraguay basins, and a discussion of the origin of the Paraguayan fauna. Proceedings of the California Academy of Science, 23:99–114.

Planquette, P., P. Keith, and P.-Y. Le Bail. 1996. Atlas des poissons d'eau douce de Guyane (Tome 1). Muséum National d'Historie Naturelle, Ministére de l'Environnement. 1–431.

Portugal, L.P.S. 1990. Revisão sistematica do gênero *Triportheus* Cope (Teleostei, Characiformes, Characidae). Unpublished Ph.D. dissertation, Instituto Biociências da São Paulo: Universidade de São Paulo. 192pp.

Reis, R.E. 1989. Systematic revision of the neotropical characid subfamily Stethaprioninae (Pisces, Characiformes). Comunicações do Museu de Ciências da PUCRS, Série Zoologia, Porto Alegre, 2:3–86.

Reis, R.E. 1997. Revision of the neotropical catfish genus *Hoplosternum* (Ostariophysi: Siluriformes: Callichthyidae), with the description of two new genera and three new species. Ichthyological Exploration of Freshwaters, 7:299–326.

Retzer, M.E. and L.M. Page. 1996. Systematics of the stick catfishes, *Farlowella* Eigenmann & Eigenmann (Pisces, Loricariidae). Proceedings of the Academy of Natural Sciences of Philadelphia, 147:33–88.

Rosa, R.S. 1985. A systematic revision of the South American freshwater stingrays (Chondrichthyes: Potamotrygonidae). Unpublished Ph.D. dissertation, Virginia: College of William and Mary. 523pp.

Schaefer, S.A. 1997. The neotropical cascudinhos: systematics and biogeography of the Otocinclus catfishes (Siluriformes: Loricariidae). Proceedings of the Academy of Natural Sciences of Philadelphia, 148:1–120.

Silfvergrip, A.M.C. 1996. A systematic revision of the neotropical catfish genus *Rhamdia* (Teleostei, Pimelodidae). Stockholm: Swedish Museum of Natural History. 156pp.

Soares-Porto, L.M. 1996. Análise filogenética dos Centromochlidae. redefinição e revisão taxonômica de *Tatia* A. de Miranda Ribeiro, 1911 (Osteichthyes, Siluriformes, Doradoidea). Unpublished Ph.D. dissertation. São Paulo: Universidade de São Paulo. 278pp.

Toledo-Piza, M. 1997. Systematic revision of the Cynodontinae (Teleostei: Ostariophysi: Characiformes). Unpublished Ph.D. dissertation. New York: The City University of New York. 394pp.

Uj, A. 1987. Les Cheirodontinae (Characidae, Ostariophysi) du Paraguay. Revue Suisse de Zoologie, 94:129–175.

Vari, R.P. 1977. Notes on the characoid subfamily Iguanodectinae, with a description of a new species. American Museum Novitates, 2612:1–6.

Vari, R.P. 1984. Systematics of the neotropical characiform genus *Potamorhina* (Pisces: Characiformes). Smithsonian Contributions to Zoology, 400:1–36.

Vari, R.P. 1989. Systematics of the neotropical characiform genus *Psectrogaster* Eigenmann and Eigenmann (Pisces: Characiformes). Smithsonian Contributions to Zoology, 481:1–43.

Vari, R.P. 1991. Systematics of the neotropical characiform genus *Steindachnerina* Fowler (Pisces: Ostariophysi). Smithsonian Contributions to Zoology, 507:1–118.

Vari, R.P. 1992. Systematics of the neotropical characiform genus *Curimatella* Eigenmann and Eigenmann (Pisces: Ostariophysi), with summary comments on the Curimatidae. Smithsonian Contributions to Zoology, 533:1–48.

Weitzman, M. 1985. *Hyphessobrycon elachys,* a new minature characid from eastern Paraguay (Pisces: Characiformes). Proceedings of the Biological Society of Washington, 98:799–808.

Weitzman, S.H. 1954. The osteology and relationships of the South American characid fishes of the subfamily Gasteropelecinae. Stanford Ichthyological Bulletin, 4:213–263.

Weitzman, S.H. 1960. Further notes on the relationships and classification of the South American characid fishes of the subfamily Gasteropelecinae. Stanford Ichthyological Bulletin, 7:217–239.

Wietzman, S.H. and S.V. Fink. 1985. Xenurobryconin phylogeny and putative pheromone pumps in glandulocaudine fishes (Teleostei: Characidae). Smithsonian Contributions to Zoology, 421:1–121.

Wietzman, S.H. and L. Palmer. 1997. A new species of *Hyphessobrycon* (Teleostei: Characidae) from the Neblina region of Venezuela and Brazil, with comments on the putative 'rosy tetra clade'. Ichthyological Exploration of Freshwaters, 7:209–242.

Zanata, A.M. 1997. *Jupiaba,* um novo género de Tetragonopterinae com osso pélvico em forma de espinho (Characidae, Characiformes). Iheringia, Série Zoologia, Porto Alegre, 83:99–136.